T0335574

VOLUME ONE HUNDRED AND SEVEN

ADVANCES IN
COMPUTERS

VOLUME ONE HUNDRED AND SEVEN

ADVANCES IN
COMPUTERS

Edited by

ATIF M. MEMON
College Park, MD,
United States

ACADEMIC PRESS

An imprint of Elsevier

Academic Press is an imprint of Elsevier
50 Hampshire Street, 5th Floor, Cambridge, MA 02139, United States
525 B Street, Suite 1800, San Diego, CA 92101-4495, United States
The Boulevard, Langford Lane, Kidlington, Oxford OX5 1GB, United Kingdom
125 London Wall, London, EC2Y 5AS, United Kingdom

First edition 2017

ISBN: 978-0-12-812228-0
ISSN: 0065-2458

For information on all Academic Press publications
visit our website at https://www.elsevier.com/books-and-journals

  Working together
to grow libraries in
developing countries

www.elsevier.com • www.bookaid.org

Publisher: Zoe Kruze
Acquisition Editor: Zoe Kruze
Editorial Project Manager: Shellie Bryant
Production Project Manager: James Selvam
Senior Cover Designer: Greg Harris

Typeset by SPi Global, India

CONTENTS

Preface   *vii*

1. Pitfalls and Countermeasures in Software Quality Measurements and Evaluations   1

Hironori Washizaki

  1. Introduction   2
  2. Pitfall: Negative Hawthorne Effects   2
  3. Pitfall: Organization Misalignment   8
  4. Pitfall: Uncertain Future   12
  5. Pitfall: Self-certified Quality   15
  6. Conclusion and Future Prospective   18
  Acknowledgments   19
  References   19
  About the Authors   21

2. Uncertainty-Wise Testing of Cyber-Physical Systems   23

Shaukat Ali, Hong Lu, Shuai Wang, Tao Yue, and Man Zhang

  1. Introduction   24
  2. Uncertainty-Wise Model-Based Testing   25
  3. Uncertainty-Wise Multiobjective Test Optimization   38
  4. The State of Art   63
  5. Conclusion and Future Research Directions   76
  Acknowledgments   78
  Appendix   78
  References   83
  About the Authors   92

3. Testing the Control-Flow, Data-Flow, and Time Aspects of Communication Systems: A Survey   95

Rachida Dssouli, Ahmed Khoumsi, Mounia Elqortobi, and Jamal Bentahar

  1. Introduction to Communication Software System Testing   96
  2. Basic Concepts of Testing   97
  3. Testing the Control and Data-Flow Aspects   102
  4. Testing the Communication Aspect   115
  5. Testing the Time Aspect   121

v

6. Discussion and Conclusion   142
Acknowledgment   143
References   143
About the Authors   154

4. Advances in Testing Software Product Lines   **157**
Hartmut Lackner and Bernd-Holger Schlingloff

1. Introduction   158
2. Specification of Variability   160
3. Model-Based Testing for Product Lines   170
4. Assessment of Product Line Test Suites   183
5. Test-Driven Product Sampling   193
6. Assignment of Product Line Test Cases   200
7. Related Work   206
8. Future Developments   209
9. Summary and Conclusion   210
References   211
About the Authors   216

5. Advances in Model-Based Testing of Graphical User Interfaces   **219**
Fevzi Belli, Mutlu Beyazıt, Christof J. Budnik, and Tugkan Tuglular

1. Introduction: User Interfaces and Their Holistic Testing   220
2. Modeling of GUIs of Interactive Systems   223
3. Testing and Test Optimization Exemplified by GUI-Modeling With ESG 241
4. Contract-Based Testing of GUIs   253
5. Rationalization and Automation of GUI Testing   262
6. Conclusions   271
References   274
About the Authors   279

PREFACE

This volume of *Advances in Computers* is the 107th in this series. This series, which has been continuously published since 1960, presents in each volume four to seven chapters describing new developments in software, hardware, or uses of computers. For each volume, I invite leaders in their respective fields of computing to contribute a chapter about recent advances.

Volume 107 focuses on five topics, all related to the important subject of software quality assurance. In Chapter 1, entitled "Pitfalls and Countermeasures in Software Quality Measurements and Evaluations," Prof. Hironori Washizaki discusses common pitfalls and their countermeasures in software quality measurements and evaluations based on research and practical achievements. The pitfalls include negative Hawthorne effects, organization misalignment, uncertain future, and self-certified quality. Corresponding countermeasures include goal–oriented multidimensional measurements, alignment visualization and exhaustive identification of rationales, prediction incorporating uncertainty and machine-learning based measurement improvement, and standard/pattern-based evaluation.

In Chapter 2, entitled "Uncertainty–Wise Testing of Cyber–Physical Systems," authors Prof. Shaukat Ali, Prof. Hong Lu, Prof. Shuai Wang, Prof. Tao Yue, and Prof. Man Zhang discuss how uncertainty-wise testing explicitly addresses known uncertainty about the behavior of a System Under Test, its operating environment, and interactions between the system and its operational environment, across all testing phases, including test design, test generation, test optimization, and test execution, with the aim to mainly achieve the following two goals. First, uncertainty–wise testing aims to ensure that the system deals with known uncertainty adequately. Second, uncertainty-wise testing should be also capable of learning new (previously unknown) uncertainties such that the system's implementation can be improved to guard against newly learned uncertainties during its operation. The necessity to integrate uncertainty in testing is becoming imperative because of the emergence of new types of intelligent and communicating software-based systems such as Cyber-Physical Systems. Intrinsically, such systems are exposed to uncertainty because of their interactions with highly indeterminate physical environments. The authors present their understanding and experience of uncertainty-wise testing from the aspects of uncertainty-wise model-based testing, uncertainty-wise modeling and evolution of test ready models, and uncertainty-wise multiobjective test optimization, in the context of testing systems under uncertainty.

In Chapter 3, entitled "Testing the Control-Flow, Data-Flow, and Time Aspects of Communication Systems: A Survey" authors Prof. Rachida Dssouli, Prof. Ahmed Khoumsi, Prof. Mounia Elqortobi, and Prof. Jamal Bentahar present a survey of testing the control-flow, data-flow, and time aspects of communication systems, which are critical national infrastructure that support users, corporations, and governments. Their description, implementation, and testing are subject to verification by standardization bodies. Formal models have been investigated to establish the conformity of communication protocols to these systems' standards over the past four decades, including Finite State Machine models for control aspects, Extended Finite State Machine models for both the control-flow and the data-flow aspects, and Timed Automata for modeling the time aspect.

In Chapter 4, "Advances in Testing Software Product Lines," authors Prof. Hartmut Lackner and Prof. Bernd-Holger Schlingloff describe recent techniques and results in model-based testing of software product lines. Presently, more and more software-based products and services are available in many different variants to choose from. However, this brings about challenges for the software quality assurance processes. Since only few of all possible variants can be tested at the developer's site, several questions arise. How shall the variability be described in order to make sure that all features are being tested? Is it better to test selected variants on a concrete level, or shall the whole software product line be tested abstractly? What is the quality of a test suite for a product line? If it is impossible to test all possible variants, which products should be selected for testing? Given a certain product, which test cases are appropriate for it, and given a test case, which products can be tested with it? The authors address these questions from an empirical software engineering perspective. They sketch modeling formalisms for software product lines. Then, they compare domain-centered and application-centered approaches to software product line testing. They define mutation operators for assessing software product line test suites. Subsequently, they analyze methods for selecting product variants on the basis of a given test suite. Finally, they show how model-checking can be used to determine whether a certain test case is applicable for a certain product variant. For all the methods, they describe supporting tools and algorithms.

In Chapter 5, "Advances in Model-Based Testing of Graphical User Interfaces," authors Prof. Fevzi Belli, Prof. Mutlu Beyazit,

Prof. Christof J. Budnik, and Prof. Tugkan Tuglular, discuss how to test software systems that contain a Graphical user interface, which enables comfortable interactions of the computer-based systems with their environment. Large systems usually require complex GUIs which are commonly fault-prone and thus are to be carefully designed, implemented, and tested. As thorough testing is not feasible, techniques are favored to test relevant features of the system under test that will be specifically modeled. This chapter summarizes, reviews, and exemplifies conventional and novel techniques for model-based GUI testing.

I hope that you find these articles of interest. If you have any suggestions of topics for future chapters, or if you wish to be considered as an author for a chapter, I can be reached at atif@cs.umd.edu.

PROF. ATIF M. MEMON, PH.D.
College Park, MD, United States

CHAPTER ONE

Pitfalls and Countermeasures in Software Quality Measurements and Evaluations

Hironori Washizaki*,†

*National Institute of Informatics, Waseda University, Tokyo, Japan
†SYSTEM INFORMATION CO., LTD., Tokyo, Japan

Contents

1. Introduction   2
2. Pitfall: Negative Hawthorne Effects   2
   2.1 Countermeasure: Goal Orientation   3
   2.2 Countermeasure: Multidimensional Measurements   7
3. Pitfall: Organization Misalignment   8
   3.1 Countermeasure: Visualization of Relationships Among Organizational
       Goals, Strategies, and Measurements   9
   3.2 Countermeasure: Exhaustive Identification of Fact-Based Rationales   11
4. Pitfall: Uncertain Future   12
   4.1 Countermeasure: Prediction Incorporating Uncertainty   13
   4.2 Countermeasure: Continuous Measurement Program Improvement by
       Machine Learning   13
5. Pitfall: Self-certified Quality   15
   5.1 Countermeasure: Standard-Based Quality Evaluation   15
   5.2 Countermeasure: Pattern-Based Quality Evaluation   17
6. Conclusion and Future Prospective   18
Acknowledgments   19
References   19
About the Authors   21

Abstract

This chapter discusses common pitfalls and their countermeasures in software quality measurements and evaluations based on research and practical achievements. The pitfalls include negative Hawthorne effects, organization misalignment, uncertain future, and self-certified quality. Corresponding countermeasures include goal-oriented multidimensional measurements, alignment visualization and exhaustive identification of rationales, prediction incorporating uncertainty and machine learning-based measurement improvement, and standard/pattern-based evaluation.

Advances in Computers, Volume 107
ISSN 0065-2458
http://dx.doi.org/10.1016/bs.adcom.2017.06.003

1

1. INTRODUCTION

Measurements to evaluate quality are essential to specify, manage, and improve the quality of product in software developments. However, a development project may become worse, such as a misleading conclusion, if the measurement program is not properly adopted. This chapter discusses common pitfalls and their countermeasures in software quality measurements and evaluations based on research and practical achievements at the Global Software Engineering Laboratory (PI: Prof. Hironori Washizaki) of Waseda University in collaboration with many software companies [1]. Table 1 summarizes the specific pitfalls addressed and their corresponding countermeasures.

2. PITFALL: NEGATIVE HAWTHORNE EFFECTS

Measurements are so powerful that they drive people's behavior. This phenomenon is known as the Hawthorne effect (or the observer effect). It was derived from famous extensive productivity research conducted at the Western Electric/AT&T Hawthorne plant between 1924 and 1932, which confirmed that whatever management paid attention to and measured improved [2].

Table 1 Pitfalls and Countermeasures in Quality Measurements and Evaluations

Pitfall	Countermeasure
Negative Hawthorne effects	Goal orientation
	Multidimensional measurements
Organization misalignment	Visualization of relationships among organizational goals, strategies, and measurements
	Exhaustive identification of rationales
Uncertain future	Prediction incorporating uncertainty
	Measurement program improvement by machine learning
Self-certified quality	Standard-based evaluation
	Pattern-based evaluation

Thus, measurements should be carefully employed in software development and quality management to help stakeholders focus on what is truly important. Otherwise, quality may improve with regard to the measurements, while quality of aspects not measured may decline at the expense of the overall quality. This is a common symptom when a measurement program is build based on available data or what is of most interest to the metrics engineer [2].

There are at least two countermeasures to prevent negative Hawthorne effects: goal orientation and multidimensional measurements. The former contributes to clarifying the focus and corresponding measurements, while the latter incorporates various aspects to ensure total quality.

2.1 Countermeasure: Goal Orientation

Goal orientation is a generic term for approaches involving goal setting and variable derivation in a top-down manner. Goal–Question–Metrics (GQM, hereafter) is a goal-oriented approach to define a measurement program from the top goal [3]. GQM takes the following three steps to define a measurement program [2]:

(1) Identify the Goal for the product/process/resource from the viewpoint of the actual "customer" of the measurement program.

(2) Determine the Questions that characterize how achievement of the goal is assessed.

(3) Define the Metrics that quantitatively answer each question.

GQM is particularly useful to capture the nature of software quality since quality is an abstract and inherently invisible concept. Applying GQM makes it much easier to focus on what is truly important for the "customer" and build a measurement program based on the goal instead of available data. Consequently, GQM may mitigate the possibility of negative Hawthorne effects and turn them into positive ones.

There are many successful cases of GQM adoption in software quality measurements, including:

- OGIS-RI Co. and GSE jointly built a static analysis and measurement tool called Adqua to evaluate the quality of embedded program source codes written in C language. Measurements in Adqua have been identified using GQM and the ISO9126-1 quality model [4]. The GQM model consists of 10 goals to evaluate quality subcharacteristics, 47 questions, 101 subquestions, and 236 metrics. Fig. 1 shows an excerpt of the

Characteristic	Subcharacteristic	Goal	Question	Subquestion	Metric
Reliability	Maturity	Purpose : Evaluate Issue : The frequency of faults Object: Source code Viewpoint: End-user	Q0100: Is the code not prone to faults?	Q0101: Has memory been initialized properly?	MF1134: Number of uninitialized const objects MF1107: Number of arrays with fewer initialization values than elements MF1133: Number of strings which do not maintain null termination MF1169: Number of enumerations not adequately initialized
				......	......
			Q0200: Is the scope not too large?	Q0201: Is the number of partition elements appropriate?	MMd027: Number of subelements MMd008: Number of functions MF1003: Effective number of lines
			Q0400: Is it possible to estimate the size of resources to be used?	Q0401: Is there not any recursive call?	Msy021: Number of recursive paths
	Fault tolerance	......	......	......	......
Maintainability	Analyzability	Purpose: Evaluate Issue : The easiness of identifying styles, structure, behavior, and parts for maintenance Object: Source code Viewpoint: Developer	Q3700: Are the functions not too complicated?	Q3701: Is the function-call nesting not deep?	MFn095: Depth of layers in call graph
				Q3702: Is the logic not too complex?	MFn066: Max. nesting depth in control structure MFn072: Cyclomatic number MFn069: Estimated no. of static paths

Fig. 1 GQM model to evaluate the quality of C programs (excerpt) [4].

model. For example, several language-independent questions (e.g., Q3700) help to evaluate how easily the source code is analyzed. Because Q3700 is quite abstract and difficult to measure directly, it is divided into several subquestions, including Q3701 and Q3702. Finally, metrics are assigned to each subquestion, allowing useful data to assess the goal to be obtained. The single metric, MFn095, is assigned to Q3701, and three metrics, MFn066, MFn072, and MFn069, are assigned to Q3702. Thus, the source code quality can be evaluated via quality sub-characteristic units from the measurement. By conducting experiments targeting several embedded programs, it has been confirmed that Adqua can be used effectively to evaluate programs for reliability, maintainability, reusability, and portability. Adqua has been used to evaluate embedded programs in Japan successfully for over 5 years.

- GSE, OGIS-RI Co., and Yamaha Corporation extended the above-mentioned tool to evaluate the reusability of C language program source code more precisely by adopting GQM to identify a set of metrics [5]. By applying the tool to 10 actual projects involving the development of existing software modifications and adoptions, it has been confirmed that these metrics effectively reflect and estimate the magnitude of necessary effort to reuse a target.

- GSE and FUJITSU CONNECTED TECHNOLOGIES investigated the impact of software transfer from one development organization to another organization on software maintainability and reliability by introducing the concept of "origins" as files' creation and modification histories [6]. They adopted GQM to specify necessary measurements to determine maintainability and reliability under the context of software transfer. Fig. 2 shows the GQM model constructed by setting goals to evaluate specific quality characteristics. Measurements are from the static analysis tool Adqua. By analyzing two open source projects, OpenOffice and VirtualBox, which were each developed by a total of three organizations, the results show that files modified by multiple organizations or developed by later organizations tend to be faultier due to the increase in complexity and modification frequency. The concept of origins as well as the measurements specified has been utilized to investigate the impact of individual developer's experience on the software quality [7,8] and to support the overall comprehension of large programs with long histories involving transfers [9]. Fig. 3 shows an example of "Origin City," which represents the measurement values for files with different origins in the form of stacked 3D buildings [9].

Goal	Question	Metric
Evaluate reliability (maturity)	Is there no unnecessary accessibility to internal elements?	Number of public methods
		Number of public attributes
	Is the memory space initialized appropriately?	Number of static objects which are not initialized explicitly
	Is the code size appropriate?	Physical lines of code
Evaluate maintainability (analyzability)	Is the hierarchical structure appropriate?	Depth of inheritance tree
	Is the abstraction appropriate?	Lack of cohesion in methods
	Are elements concealed appropriately?	Number of global variables
		Number of public static attributes
Evaluate maintainability (changeability)	Are there no complex sentences?	Number of lines with multiple statements
		Rate of methods which call methods in other classes
		Number of methods in other classes which this class calls
	Are effects of external changes limited?	Number of functions using global variables defined in other files
		Number of methods using public static attributes defined in other files
		Number of functions and methods defined in other files which this file calls
Evaluate maintainability (stability)		Number of global variables defined in other files which this file uses
		Number of global variables used in other files
	Are effects of changes on the outside limited?	Number of public static attributes used in other files
		Number of functions and methods defined in other files which call functions/methods defined in this file

Fig. 2 GQM model to evaluate the reliability and maintainability of programs [6].

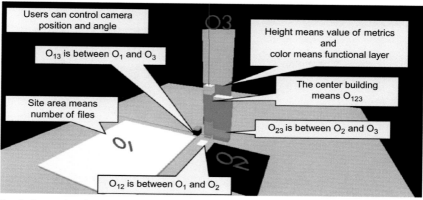

Fig. 3 Example of Origin City [9].

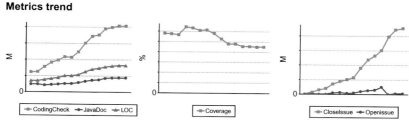

Fig. 4 Dashboard visualizing multiple measurement results (excerpt) [10].

2.2 Countermeasure: Multidimensional Measurements

Beside goal-oriented measurements, it is also important to measure and evaluate targets multidimensionally to cover various aspects and ensure total quality since any feature may have side effects or unintended quality characteristics. A typical example is the trade-off between maintainability and performance (i.e., time behavior); a program tuned for computing performance may be less comprehensible for human developers.

Multidimensional measurements and evaluations are particularly crucial to grasp the total quality of software. For example in Ref. [4], the GQM model and specified measurements successfully cover most major quality characteristics to measure and evaluate embedded C programs multidimensionally.

A multidimensional evaluation may reveal trends and tendencies of software quality in detail. For example, GSE and Yahoo Japan jointly built a dashboard (Fig. 4) to visualize multiple measurement results based on the

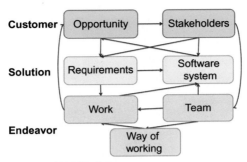

Fig. 5 Relationships among SEMAT alphas.

underlying GQM model to support decision making [10]. Visualizing the multidimensional measurement results allows users to easily grasp possible side effects and the overall total quality.

A multidimensional evaluation is also useful to capture the software development process and project status. For example, the SEMAT (Software Engineering Methods and Theory) initiative proposed a framework, the SEMAT Kernel, to reason about the progress of stakeholders and the health of their endeavors in terms of six different but mutually dependent concerns: opportunity, stakeholders, requirements, software system, work, team, and way of working [11]. These concerns are called "alphas." Alphas are essential elements of a software engineering endeavor, and their progress and health must be assessed. Fig. 5 shows the relationships among these alphas. Through this framework, stakeholders can capture a project's status multidimensionally rather than through work products (such as documents).

In cooperation with the SEMAT Japan Chapter, a working group of ITA (Information Technology Alliance, which is an association of Japanese information technology companies) analyzed existing project failure cases using the SEMAT Kernel [12]. They then identified root causes and countermeasures of these cases efficiently from wider viewpoints. Fig. 6 shows an example of root cause analysis results by analyzing a failure case through the relationships among alphas such as opportunity and stakeholders [13].

3. PITFALL: ORGANIZATION MISALIGNMENT

A measurement program must be fully aligned with organizational goals and strategies; otherwise, even if GQM is adopted to clarify measurement goals and corresponding metrics, these goals and metrics may be useless

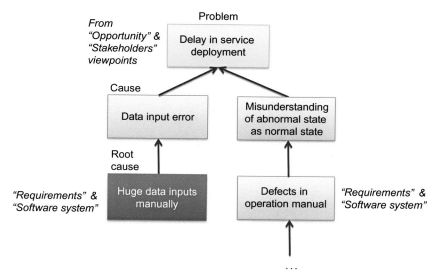

Fig. 6 Root cause analysis based on SEMAT alphas [13].

from the organizational management's point of view since their contributions to the organization may be unclear without coherent rationales. To prevent such misalignments, at least two countermeasures are possible: visualization of relationships among organizational goals, strategies, and measurements and exhaustive identification of fact-based rationales.

3.1 Countermeasure: Visualization of Relationships Among Organizational Goals, Strategies, and Measurements

By visualizing the relationships among organizational units, goals, strategies, and measurements, whether (or not) the measurement program is consistent and fully aligned with the organization becomes clear. GQM + Strategies, which was developed by Basili *et al.*, is an extension of GQM that aligns and assesses the organizational and business goals at each organizational level to the overall strategies and goals of the organization [14,15]. Fig. 7 shows the structure of the GQM + Strategies model (called grid) [16]. GQM + Strategies has been used to establish management strategies and plans, determine the value of a contribution, ensure the integrity of a goal between a purchaser and a contractor, and evaluate management based on quantitative data.

There are many successful cases applying GQM + Strategies with extensions for measurement-based IT business alignment, including:

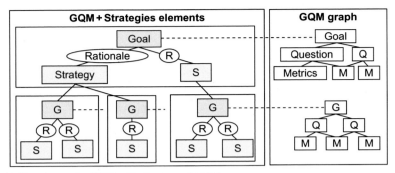

Fig. 7 Structure of a GQM + Strategies grid [16].

- Recruit Sumai Company Co., which provides services and products related to housing, adopted GQM + Strategies [16]. In this case, GQM + Strategies maintains consistency within a vertical refinement tree composed by goals, strategies, and measurements. In addition, since horizontal relations such as conflicting ones at different branches (Fig. 8) may be missed in the original GQM + Strategies approach, GSE proposed the Horizontal Relation Identification Method (HoRIM) to identify horizontal relations by employing Interpretive Structural Modeling (ISM) [16,17]. Applying GQM + Strategies along with HoRIM identifies about 1.5 times more horizontal relations than an ad hoc review.
- GSE together with Yahoo Japan proposed a method, GO-MUC method (Goal-oriented Measurement for Usability and Conflict) (Fig. 9), which is a goal–oriented strategy design approach considering the requirements of both the user and the business by combining GQM + Strategies and Persona approaches [18]. Applying GO-MUC to an actual software service development and operation demonstrated that GO-MUC can identify the interest between the business side and users side, realizing more effective and user-friendly strategies to resolve conflicting interests.
- SYSTEM INFORMATION Co., Ltd. has achieved CMMI Capability Level 5, which is the highest maturity level.[a] The adoption of GQM + Strategies has contributed to the achievement by aligning organizational goals, strategies, and measurements over organizational units. In their consulting services, the application of GQM + Strategies is recommended when customers implement CMMI High Maturity practices.

---

[a] http://www.sysj.co.jp/endeavor/endeavor01.html.

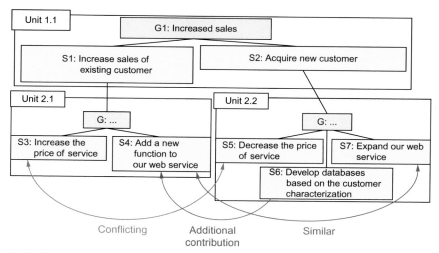

Fig. 8 Horizontal relations in GQM+Strategies grid [16].

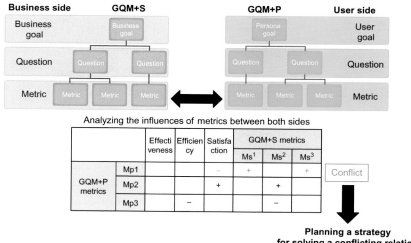

Fig. 9 Overview of GO-MUC [18].

3.2 Countermeasure: Exhaustive Identification of Fact-Based Rationales

GQM + Strategies extracts strategies from goals based on rationales such as fact-based contexts and assumptions. A lack of rationales tends to be misleading and may result in deriving incorrect strategies. Consequently, rationales must be identified exhaustively.

	Order reception Grp.	Shipment Grp.	Inventory control Grp.	...	TBD
Actor / Viewpoint					
Order reception Grp.	C1:··· C2:···		C3:Inventory control group sometimes mistake the number of the stocks.		C4:No one integrates complaints from customers in customer service
Shipment Grp.		C5:··· A1:···	A2:···		
Inventory control Grp.					
...					
TBD					

Fig. 10 Context–Assumption–Matrix [19].

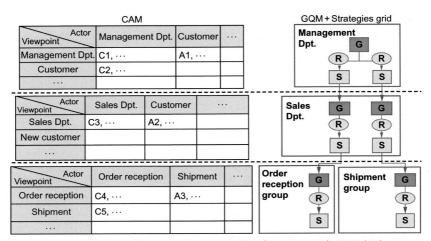

Fig. 11 Iterative process of GQM+Strategies grid refinement with CAM [19].

GSE proposed a method named CAM (Context–Assumption–Matrix) to extract contexts and assumptions efficiently and exhaustively by analyzing the relationships between stakeholders. Fig. 10 shows an example of CAM [19]. CAM organizes common contexts and assumptions between stakeholders into a two-dimensional table. CAM can be employed as part of the GQM+Strategies grid construction process to refine business goals and strategies iteratively from top to bottom (Fig. 11).

4. PITFALL: UNCERTAIN FUTURE

Quality measurements and evaluations are often conducted based on the strong assumption that the "future is an extension of the present." Especially in an era of uncertainty in computing and environments, there is no guarantee that a prediction or estimation of quality or related elements based on past data at a certain time point will be correct in the future. At least two countermeasures will prevent such incorrect predictions and estimations: quality prediction in consideration with uncertainty and continuous improvement of quality measurement and continuous measurement program improvement by machine learning.

4.1 Countermeasure: Prediction Incorporating Uncertainty

Among the various quality characteristics, software reliability is a critical component of computer system availability. Software reliability growth models (SRGMs), such as the Times Between Failures Model and Failure Count Model, can indicate whether a sufficient number of faults have been removed to release the software [20]. Although logistic and Gompertz curves are both well-known software reliability growth curves, neither can account for the dynamics of software development because developments are affected by various elements in the development environment (e.g., skills of the development team and changing requirements).

To adapt to changes, GSE proposed a generalized software reliability model (GSRM) based on a stochastic process to simulate developments, which include uncertainties and dynamics such as unpredictable changes in the requirements and the number of team members [20]. Fig. 12 shows combinations of three different types of dynamics and three different types of uncertainty, resulting in nine models. Fig. 13 plots the ratio of the cumulative number of detected faults at time t vs the total number of detected faults for the entire project using these nine models, where the x-axis represents time in arbitrary units.

GSE assessed two actual datasets using our formulated equations, which are related to three types of development uncertainties by employing simple approximations in GSRM. The results confirm that developments can be evaluated quantitatively. Additionally, a comparison of GSRM with existing software reliability models demonstrates that the approximation by GSRM is more precise than those by existing models [20].

4.2 Countermeasure: Continuous Measurement Program Improvement by Machine Learning

Employing automated measurement tools and thresholds allows quality evaluations to be conducted automatically. For example, program static analysis tools can be used to measure attributes of programs. The

The number of detected faults per unit time is constant, but the uncertainty increases near the end. This model is similar to a logistic curve. (Model 1-1)	The number of detected faults per unit time and uncertainty are constant. (Model 1-2)	The number of detected faults per unit time is constant, but the uncertainty decreases over time (e.g., the team matures over time). (Model 1-3)
The number of detected faults per unit time changes at t_1, and the uncertainty increases near the end (e.g., new members join the project at time t_1). (Model 2-1)	The number of detected faults per unit time changes at t_1, but the uncertainty is constant. (Model 2-2)	The number of detected faults per unit time changes at t_1, but the uncertainty decreases over time. (Model 2-3)
The number of detected faults per unit time and the uncertainty increase near the end (e.g., increasing manpower with time). (Model 3-1)	The number of detected faults per unit time increases, but the uncertainty is constant. (Model 3-2)	The number of detected faults per unit time increases, but the uncertainty decreases over time. (Model 3-3)

Fig. 12 Combinations of dynamics and uncertainty in the generalized software reliability model [20].

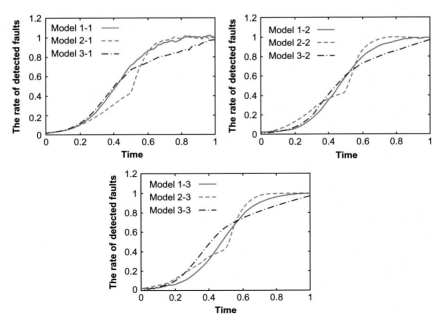

Fig. 13 Example plots of the generalized software reliability model [20].

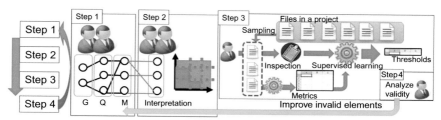

Fig. 14 Iterative process to improve a measurement program by machine learning [21].

measurement results can indicate high-risk program modules against given thresholds. However, the problem is how to establish appropriate measurements and thresholds from the viewpoint of quality evaluation because these may vary over time.

To solve this problem, GSE proposed a GQM-based process to search for optimal thresholds by supervised machine learning using measurement values taken from modules sampled as training data and improving measurement methods by experts' analysis of the evaluation results based on the thresholds [21]. Fig. 14 shows the entire process. Implementing the process iteratively can continuously improve and adapt a measurement program to the given context. A case study at a construction machinery company with the goal of detecting high-risk C++ files in embedded systems confirms that the proposed process is useful to achieve a goal in an iterative manner [21].

5. PITFALL: SELF-CERTIFIED QUALITY

Although quality requirements and evaluation methods can be defined for each development and operational context, these definitions should be "reasonable" from the viewpoint of targeted domains and industries. Otherwise, self-declared or self-certified software quality management without consideration of outside standards or industrial de-facts may cause the quality to decline. Such management may result in having relatively poor quality requirements or incomprehensive evaluations. At least two possible countermeasures exist: standard-based quality evaluation and pattern-based quality evaluation.

5.1 Countermeasure: Standard-Based Quality Evaluation

Several works have strived to identify software quality, but the quality of software products is often not comprehensively, specifically, or effectively defined because previous approaches focused on specific quality aspects.

Moreover, the evaluation results of quality metrics often depend on software stakeholders, making it difficult to compare quality evaluation results across software products. ISO/IEC has tried to define evaluation methods for the quality of software products and provide common standards, called the SQuaRE (Systems and software Quality Requirements and Evaluation) series [22,23], including ISO/IEC 25022:2016 [24] and ISO/IEC 25023:2016 [25]. Because the SQuaRE series includes ambiguous metrics, applying the series to products and comparing results can be challenging. Thus, GSE proposed a SQuaRE-based software quality evaluation framework, which successfully concretized many product metrics and quality-in-use metrics originally defined in the SQuaRE series [26].

Fig. 15 overviews the framework. The framework is composed of two parts: "Product Quality" and "Quality in Use." The former contains internal and external product quality characteristics and metrics based on ISO/IEC 25023:2016, whereas the latter has quality characteristics and quality-in-use metrics based on ISO/IEC 25022:2016. Since the product quality is expected to influence the quality in use, the framework measures both qualities to clarify the relationship. In relation to PSQ Certification System [27], GSE selected and concretized 47 product metrics and 18 quality-in-use metrics. These metrics can be concretely applied to almost any software package/service products. These metrics cover more than 50% of the metrics originally defined in the SQuaRE series.

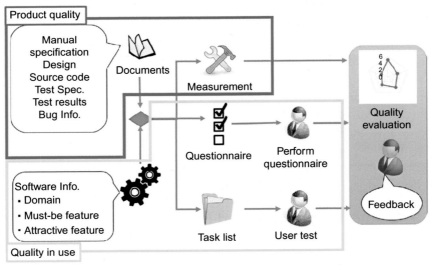

Fig. 15 Overview of SQuaRE-based comprehensive evaluation framework [26].

This framework requires manual specifications, test specifications, and bug information to measure product quality (Fig. 15). Moreover, the framework requires information collected via a questionnaire and a user test to measure quality-in-use metrics. Finally, the overall software quality is assessed based on the results to clarify what quality characteristics are sufficient (or insufficient).

Through a case study targeting a commercial software product, GSE confirmed that the framework is concretely applicable to the software package/service product. The framework together with the measurement results of 21 Japanese software products is available as the Waseda Software Quality Benchmark (WSQB) at Ref. [28].

5.2 Countermeasure: Pattern-Based Quality Evaluation

Some specific quality characteristics, especially security, are difficult to accommodate because not all software engineers are specialists in these characteristics. Patterns are reusable packages that encapsulate expert knowledge. Specifically, a pattern represents a frequently recurring structure, behavior, activity, process, or "thing" during the software development process. Many security patterns have been proposed [29,30].

For example, the Role-based Access Control (RBAC) pattern (Fig. 16), which is a representative pattern for access control, describes how to assign precise access rights to roles in an environment where access to computing resources must be controlled to preserve confidentiality and the availability requirements.

Security design patterns are difficult to implement because they are currently abstract descriptions. Additionally, validating security design patterns in the implementation phase is challenging because an adequate test case is required. Hence, a security design pattern can be inappropriately applied, which leads to serious vulnerability issues [31].

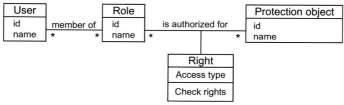

Fig. 16 Essential structure of the RBAC pattern [31].

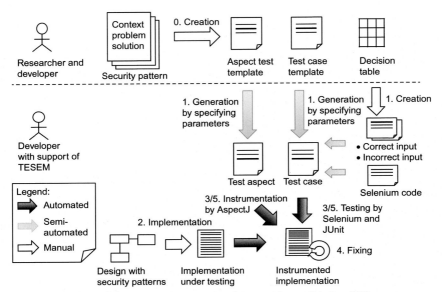

Fig. 17 Overview of the security pattern implementation validation [31].

To evaluate program implementations in terms of security, GSE proposed a method to support the implementation of security design patterns using a test template (Fig. 17). The method creates a test template from a security design pattern, which consists of an "aspect test template" to observe internal processing and a "test case template." By providing design information in the test template, a test is created to evaluate the system in the early implementation stage using a tool and fixing the code. The test can be executed repeatedly. Thus, it can validate whether a security design pattern is appropriately applied in the implementation phase. The method has been used in a previous design-level validation tool called TESEM [32,33].

6. CONCLUSION AND FUTURE PROSPECTIVE

Four pitfalls and eight corresponding countermeasures in software quality measurements and evaluations are explained using actual case studies and adaptation results mostly taken at GSE with industrial collaborations. These pitfalls and countermeasure could be utilized for efficient and effective software quality measurements and evaluations.

As future prospects, there are three major directions in quality measurement research and practice.

First, total organizational quality managements involving product, process, and resource quality measurements with their correlation analysis are expected; such truly multidimensional measurements and analysis over different targets should be beneficial for avoiding the negative Hawthorne effects and the organizational misalignment.

Second, quality measurements for recent computing platforms such as AI-based strategic software systems and IoT/cloud/edge software systems are needed to be investigated. These measurements would incorporate the uncertain future since these systems are expected to be adaptive and evolvable.

Finally, worldwide quality benchmarking of actual software products, processes, and resources is expected for promoting independent quality certifications. Toward this direction, GSE published the WSQB including product quality data of 21 Japanese software; we expect that the benchmark with its underlying framework will support promotion of quality benchmarking movement globally.

ACKNOWLEDGMENTS

Achievements of GSE introduced in this chapter were supported by JSPS KAKENHI Grant Numbers 25330091 and 16H02804, "Research Initiative on Advanced Software Engineering in 2015" supported by Software Reliability Enhancement Center (SEC), and IISF SSR Forum 2015 and 2016.

REFERENCES

[1] Global Software Engineering Laboratory, Waseda University, https://www.waseda.jp/inst/gcs/en/labo/globalsoftware/.
[2] L.M. Laird, M.C. Brennan, Software Measurement and Estimation: A Practical Approach, Wiley-IEEE Computer Society, Los Alamitos, CA, 2006.
[3] V.R. Basili, G. Caldiera, D. Rombach, R. van Solingen, Goal Question Metric (GQM) approach, Encyclopedia of Software Engineering, John Wiley & Sons, Hoboken, NJ, 2002.
[4] H. Washizaki, R. Namiki, T. Fukuoka, Y. Harada, H. Watanabe, in: A framework for measuring and evaluating program source code quality, Proceedings of the 8th International Conference on Product Focused Software Development and Process Improvement (PROFES 2007), Springer LNCS, 2007, pp. 284–299.
[5] H. Washizaki, T. Koike, R. Namiki, H. Tanabe, in: Reusability metrics for program source code written in C language and their evaluation, Proceedings of the 13th International Conference on Product-Focused Software Development and Process Improvement (PROFES 2012), pp. 89–103, June 13–15, 2012, 2012.
[6] S. Sato, H. Washizaki, Y. Fukazawa, S. Inoue, H. Ono, Y. Hanai, M. Yamamoto, in: Effects of organizational changes on product metrics and defects, Proceedings of the 20th Asia-Pacific Software Engineering Conference (APSEC 2013), pp. 132–139, Bangkok, Thailand, December 2–5, 2013, 2013.

[7] R. Ando, S. Sato, C. Uchida, H. Washizaki, Y. Fukazawa, S. Inoue, H. Ono, Y. Hanai, M. Kanazawa, K. Sone, K. Namba, M. Yamamoto, in: How does defect removal activity of developer vary with development experience?, Proceedings of the 27th International Conference on Software Engineering and Knowledge Engineering (SEKE 2015), Wyndham Pittsburgh University Center, Pittsburgh, PA, July 6–8, 2015, 2015.

[8] T. Tsunoda, H. Washizaki, F. Yosiaki, I. Sakae, Y. Hanai, M. Kanazawa, in: Evaluating the work of experienced and inexperienced developers considering work difficulty in software development, 18th IEEE/ACIS International Conference on Software Engineering, Artificial Intelligence, Networking and Parallel/Distributed Computing, June 26–28, 2017, Kanazawa, Japan, 2017.

[9] R. Ishizue, H. Washizaki, Y. Fukazawa, S. Inoue, Y. Hanai, M. Kanazawa, K. Namba, in: Metrics visualization technique based on the origins and function layers for OSS-based development, 4th IEEE Working Conference on Software Visualization (VISSOFT 2016), NIER Track, Raleigh, NC, October 3–4, 2016, 2016.

[10] H. Nakai, K. Honda, H. Washizaki, Y. Fukazawa, K. Asoh, K. Takahashi, K. Ogawa, M. Mori, T. Hino, Y. Hayakawa, Y. Tanaka, S. Yamada, D. Miyazaki, in: Initial industrial experience of GQM-based product-focused project monitoring with trend patterns, 21st Asia-Pacific Software Engineering Conference (APSEC 2014), Jeju, Korea, December 1–4, 2014, 2014.

[11] I. Jacobson, P.-W. Ng, P.E. McMahon, I. Spence, S. Lidman, The essence of software engineering: the SEMAT Kernel, Commun. ACM 55 (12) (2012) 42–49.

[12] Information Technology Alliance, http://ita.gr.jp/.

[13] H. Washizaki, in: Analyzing and refining project failure cases from wider viewpoints by using SEMAT essence, Essence Conference in Seoul, 2017.

[14] V. Basili, M. Lindvall, M. Regardie, C. Seaman, J. Heidrich, J. Munch, D. Rombach, A. Trendowicz, Linking software development and business strategy through measurement, Computer 43 (4) (2010) 57–65.

[15] V. Basili, A. Trendowicz, M. Kowalczyk, J. Heidrich, C. Seaman, J. Münch, D. Rombach, Aligning Organizations Through Measurement—The GQM + Strategies Approach, Springer-Verlag, Berlin, Germany, 2014.

[16] Y. Aoki, T. Kobori, H. Washizaki, Y. Fukazawa, in: Identifying misalignment of goal and strategies across organizational units by interpretive structural modeling, Proceedings of the 49th Hawaii International Conference on System Sciences (HICSS-49), January 5–8, 2016, Grand Hyatt, Kauai, 2016.

[17] Y. Aoki, H. Washizaki, C. Shimura, Y. Senzaki, Y. Fukazawa, in: Experimental evaluation of HoRIM to improve business strategy models, 16th IEEE/ACIS International Conference on Computer and Information Science (ICIS 2017), May 24–26, 2017, Wuhan, China, 2017.

[18] C. Uchida, K. Honda, H. Washizaki, Y. Fukazawa, K. Ogawa, T. Yagi, M. Ishigaki, M. Nakagawa, in: GO-MUC: a strategy design method considering requirements of user and business by goal-oriented measurement, 9th International Workshop on Cooperative and Human Aspects of Software Engineering (CHASE 2016), collocated with ICSE 2016, Austin, TX, May 16, 2016, 2016.

[19] T. Kobori, H. Washizaki, Y. Fukazawa, D. Hirabayashi, K. Shintani, Y. Okazaki, Y. Kikushima, Exhaustive and efficient identification of rationales using GQM + Strategies with stakeholder relationship analysis, IEICE Trans. Inf. Syst. E99-D (9) (2016) 2219–2228.

[20] K. Honda, H. Washizaki, Y. Fukazawa, Generalized software reliability model considering uncertainty and dynamics: model and applications, Int. J. Softw. Eng. Knowl. Eng. (2017) 1–29 (to appear).

[21] N. Tsuda, M. Takada, H. Washizaki, Y. Fukazawa, S. Sugimura, Y. Yasuda, M. Futakami, in: Iterative process to improve GQM models with metrics thresholds

to detect high-risk files, IEEE TENCON 2016, Marina Bay Sands, Singapore, November 22–25, 2016, 2016.

[22] ISO/IEC, ISO/IEC 25000:2014 Systems and Software Engineering—Systems and Software Quality Requirements and Evaluation (SQuaRE)—Guide to SQuaRE, 2014.

[23] ISO/IEC, ISO/IEC 25010:2011 Systems and Software Engineering—Systems and Software Quality Requirements and Evaluation (SQuaRE)—System and Software Quality Models, 2011.

[24] ISO/IEC, ISO/IEC 25022:2016 Systems and Software Engineering—Systems and Software Quality Requirements and Evaluation (SQuaRE)—Measurement of Quality in Use, 2015.

[25] ISO/IEC, ISO/IEC 25023:2016 Systems and Software Engineering—Systems and Software Quality Requirements and Evaluation (SQuaRE)—Measurement of System and Software Product Quality, 2015.

[26] H. Nakai, N. Tsuda, K. Honda, H. Washizaki, Y. Fukazawa, in: A SQuaRE-based software quality evaluation framework and its case study, IEEE TENCON 2016, Marina Bay Sands, Singapore, November 22–25, 2016, 2016.

[27] CSAJ, PSQ Certification System, http://www.psq-japan.com/english/.

[28] Global Software Engineering Laboratory, Waseda Software Quality Benchmark, http://www.washi.cs.waseda.ac.jp/?page_id=3910.

[29] N. Yoshioka, H. Washizaki, K. Maruyama, A survey on security patterns, Prog. Inform. 5 (2008) 35–47.

[30] E.B. Fernandez, H. Washizaki, N. Yoshioka, in: Abstract security patterns, Proceedings of the 15th Conference on Pattern Languages of Programs, 2008.

[31] M. Yoshizawa, H. Washizaki, Y. Fukazawa, T. Okubo, H. Kaiya, N. Yoshioka, Implementation support of security design patterns using test templates, Information 7 (2) (2016) 34.

[32] T. Kobashi, N. Yoshioka, H. Kaiya, H. Washizaki, T. Okubo, Y. Fukazawa, Validating security design pattern applications by testing design models, Int. J. Secur. Softw. Eng. 5 (4) (2014) 1–30.

[33] T. Kobashi, M. Yoshizawa, H. Washizaki, Y. Fukazawa, N. Yoshioka, H. Kaiya, T. Okubo, in: TESEM: a tool for verifying security design pattern applications by model testing, Proceedings of the 8th IEEE International Conference on Software Testing, Verification, and Validation (ICST 2015), Tool Track, April 13–17, 2015, Graz, Austria, 2015.

ABOUT THE AUTHORS

Hironori Washizaki is the Director and a Professor with the Global Software Engineering Laboratory, Waseda University, Japan. He is also a Visiting Professor with the National Institute of Informatics, and an Outside Director with the SYSTEM INFORMATION CO., LTD. He received the Ph.D. degree in information and computer science from Waseda University, in 2003. He was a Visiting Professor with the Ecole Polytechnique de Montreal, in 2015. He has long-term experience of researching and

practicing software design, reuse, quality assurance, and education. He has
served on the organizing committees of various international software engineering conferences, including ASE, ICST, SPLC, CSEE&T, SEKE, BICT,
APSEC, and AsianPLoP, on the Editorial Board of international journals,
including the Int. J. Soft. Eng. & Know. Eng., the IEICE Trans. Info. &
Sys., and the Heliyon. He has served at many professional societies, such
as the IEEE Computer Society Japan Chapter Chair, the SEMAT Japan
Chapter Chair, the IPSJ SamurAI Coding Director, ISO/IEC/JTC1
SC7/WG20 Convener, the Int. J. Agile and Extreme Software Develop-
ment Editor-in-Chief, and the IEEE Computer Society Membership at
Large for the Professional and Educational Activities Board.

CHAPTER TWO

Uncertainty-Wise Testing of Cyber-Physical Systems

Shaukat Ali*, Hong Lu*, Shuai Wang*, Tao Yue*,†, Man Zhang*
*Simula Research Laboratory, Oslo, Norway
†University of Oslo, Oslo, Norway

Contents

1. Introduction   24
2. Uncertainty-Wise Model-Based Testing   25
   2.1 Overview of Test Generation   27
   2.2 Uncertainty-Wise Test Generation   28
   2.3 Tool Support   36
   2.4 Standardization   36
3. Uncertainty-Wise Multiobjective Test Optimization   38
   3.1 Multiobjective Search Problem   39
   3.2 Uncertainty-Wise Multiobjective Search Problem   40
   3.3 Uncertainty-Wise Multiobjective Test Optimization   41
   3.4 Uncertainty-Wise Multiobjective Test Set Minimization   43
   3.5 Uncertainty-Wise Multiobjective Test Case Prioritization   45
   3.6 Examples of UWTM and UWTP   47
   3.7 Examples of Cost and Effectiveness Measures   51
   3.8 Analyzing Results   51
4. The State of Art   63
   4.1 Summary of the Systematic Mapping   66
   4.2 Multiobjective Test Optimization   73
5. Conclusion and Future Research Directions   76
Acknowledgments   78
Appendix   78
References   83
About the Authors   92

Abstract

As compared with classical software/system testing, uncertainty-wise testing explicitly addresses known uncertainty about the behavior of a System Under Test (SUT), its operating environment, and interactions between the SUT and its operational environment, across all testing phases, including test design, test generation, test optimization, and test execution, with the aim to mainly achieve the following two goals. First, uncertainty-wise testing aims to ensure that the SUT deals with known uncertainty

Advances in Computers, Volume 107
ISSN 0065-2458
http://dx.doi.org/10.1016/bs.adcom.2017.06.001

adequately. Second, uncertainty-wise testing should be also capable of learning new (previously unknown) uncertainties such that the SUT's implementation can be improved to guard against newly learned uncertainties during its operation. The necessity to integrate uncertainty in testing is becoming imperative because of the emergence of new types of intelligent and communicating software-based systems such as Cyber-Physical Systems (CPSs). Intrinsically, such systems are exposed to uncertainty because of their interactions with highly indeterminate physical environments. In this chapter, we provide our understanding and experience of uncertainty-wise testing from the aspects of uncertainty-wise model-based testing, uncertainty-wise modeling and evolution of test ready models, and uncertainty-wise multiobjective test optimization, in the context of testing CPSs under uncertainty. Furthermore, we present our vision about this new testing paradigm and its plausible future research directions.

1. INTRODUCTION

Uncertainty is unpreventable in the behavior of a Cyber–Physical Systems (CPSs) given its close interaction with its physical environment [1–4]. Predicting the exact behavior of the physical environment of a CPS is not viable, and a common practice is to make assumptions about the physical environment during the design and testing of the CPS. The correct behavior of a CPS is only guaranteed when such assumptions prove to be true. Given the complexity of problems being solved by CPSs in critical domains, these systems must function safely even when experiencing uncertainty in their physical environment to avert any harms. Thus, we argue that uncertainty (i.e., lack of knowledge) in the behavior of a CPS, its operating environment, and in their interactions must be considered explicitly during the testing phase. To this end, we propose, in this chapter, *Uncertainty-wise Testing (UWT)*, which is a new testing paradigm that explicitly takes uncertainty into consideration at various testing phases, including test design, test generation, test optimization, and test execution, to make sure that the implementation of a CPS is sufficiently robust against its uncertain physical environment.

This chapter introduces *UWT* particularly in the context of CPSs. Note that this chapter surveys the existing works in this area rather than providing new scientific and research contributions; therefore, we refer to relevant references where further details can be consulted. In particular, we survey the following three key topics in the field of *UWT*.

First, we introduce *Uncertainty-wise Model-Based Testing (UWMBT)* that focuses on model-based testing of CPSs in the presence of uncertainty with the explicit consideration of known uncertainty on test ready models. Such test ready models capture the behavior of a CPS together with uncertainty in its environment and interactions between them. Particularly, we will introduce uncertainty-wise test modeling and evolution, which aims to improve the quality of modeled test ready models with uncertainty using the real operational data of CPSs. With machine learning techniques, invariants and observed uncertainties can be abstracted from such data that can be used to further enhance test ready models. Such improved test ready models can be used to generate additional test cases as compared to the initial test ready models. In addition, uncertainty-wise test generation will also be discussed. Notice that such test ready models are the key inputs for an uncertainty-wise model-based test generation tool that generates test cases from the models using uncertainty-wise test case generation strategies. The generated test cases are then executed on a CPS to ensure that it handles the known uncertainty specified in the models during its operation, in addition to discovering unknown uncertainty, i.e., the uncertainty not specified in the models.

Second, we will survey about uncertainty-wise multiobjective test optimization. Since test case execution on a real CPS is not only time consuming but also resource intensive, test optimization is necessary. When performing multiobjective test optimization, in addition to cost and effectiveness objectives uncertainty-wise optimization objectives must also be considered. In this topic, we particularly focus on uncertainty-wise test set minimization and uncertainty-wise test case prioritization. Finally, we also provide a detailed survey of existing uncertainty-wise testing techniques for CPSs.

We organize our chapter as follows. Section 2 discusses uncertainty-wise model-based testing and uncertainty-wise test modeling and evolution in detail followed by uncertainty-wise multiobjective test optimization (Section 3). Section 4 discusses the detailed state of the art related to CPS testing under uncertainty. Finally, Section 5 summarizes and concludes the chapter. To make the chapter more readable, we provide a detailed list of abbreviations that will be used in this chapter as shown in Table 1.

2. UNCERTAINTY-WISE MODEL-BASED TESTING

In this section, we first briefly summarize two main research trends of test generation and point readers to existing literature reviews or surveys for details in Section 2.1. Second, we discuss our perspective and vision of uncertainty-wise test generation, with a particular focus on uncertainty-wise MBT (Section 2.2). Last, we highlight two important aspects (i.e., tool support in Section 2.3 and standardization in Section 2.4) in terms of enhancing the possibility of being applied in practical and industrial settings.

Table 1 List of Abbreviations

Abbreviation	Description
APML	All Paths with Maximum Length
ASP	All Simple Paths
BTRM	Belief Test Ready Model
CPS	Cyber-Physical System
ETSI	European Telecommunications Standards Institute
ICT	Information and Communications Technology
IEC	International Electrotechnical Commission
IEEE	Institute of Electrical and Electronics Engineers
ISO	International Organization for Standardization
ITU	International Telecommunication Union
JCGM	Joint Committee for Guides in Metrology
MARTE	Modeling and Analysis of Real-Time Embedded Systems
MBE	Model-Based Engineering
MBT	Model-Based Testing
MDE	Model-Driven Engineering
NIST	The National Institute of Standards and Technology
OASIS	Organization for the Advancement of Structured Information Standards
OCL	Object Constraint Language
OMA	Open Mobile Alliance
OMG	Object Management Group
OSI	Open Systems Interconnection
RFP	Request For Proposal
RTF	Revision Task Force
SACM	Structured Assurance Case Metamodel
SBT	Search-Based Testing
TDL	Test Description Language
TRM	Test Ready Model

Table 1 List of Abbreviations—cont'd

Abbreviation	Description
UMF	Uncertainty Modeling Framework
UML	Unified Modeling Language
UTF	Uncertainty Testing Framework
UTP	UML Testing Profile
UUP	UML Uncertainty Profile
UWMBT	Uncertainty-Wise Model-Based Testing
UWT	Uncertainty-Wise Testing
UWTG	Uncertainty-Wise Test Generation
UWTM	Uncertainty-Wise Multiobjective Test Set Minimization
UWTP	Uncertainty-Wise Multiobjective Test Case Prioritization

2.1 Overview of Test Generation

Automated test generation [5–8] is in principle about automatically generating test cases by following certain generation strategies, including automatically generating test data, based on one or more software artifacts such as program structures and software/system design models. Depending on which type(s) of software artifacts to use, there exist many test generation techniques such as symbolic execution [9,10], program structural coverage-based test generation [11,12], and model-based test case generation [13]. Depending on which mechanism to use for test generation, there exist other test generation techniques such as combinatorial testing [6] and search-based testing (SBT) [7,14].

When particularly looking into the research stream of MBT [15], there are many MBT techniques that have been proposed. Such techniques take very diverse types of models as input and generate test cases either automatically or semiautomatically. We term such models as Test Ready Models (TRMs) in general, which refer to all types of models that are used as input for generating test cases. TRMs can be very different, depending on testing contexts and objectives. For example, in Ref. [16], the authors proposed a restricted natural language-based test case specification language (named as RTCM) to specify test case specifications (one type of TRMs), from which executable test cases can be generated. TRMs are also often specified as UML models. For example, the authors of Ref. [17] proposed a model-based framework, named as TRUST, for automatically generating executable test cases from UML state machines.

SBT [7,8] is getting more and more attention these days. SBT is about using metaheuristic optimization search techniques (e.g., Genetic Algorithm [18]) for enabling fully automated or partially automated testing tasks such as generating test cases or test data. Researchers and practitioners have advanced this field by proposing varying types of testing techniques, summarized in Refs. [7,14]. When talking about search-based test generation particularly, interesting solutions have been proposed in the last 10 years. For example, EvoSuite [19] is an automated test suite generation tool for testing Java programs.

2.2 Uncertainty-Wise Test Generation

Considering the fact that models are the key artifacts of any MBT technique, in this section, we first put our focus on modeling methodologies. Second, we discuss uncertainty-wise test generation strategies.

2.2.1 Uncertainty-Wise Modeling

In this section, we first define TRMs in the context of MBT, based on which we further define TRMs with uncertainty information explicitly captured (Section 2.2.1.1). After that, we discuss uncertainty modeling from a general perspective (Section 2.2.1.2). In the last three subsections, we discuss modeling notations and methodologies for modeling TRMs with uncertainty (Section 2.2.1.3), how they fit into a typical system development life cycle (Section 2.2.1.4), and what measures one should take to ensure the quality of developed TRMs with uncertainty via verification and model evolution techniques (Section 2.2.1.5).

2.2.1.1 Belief Test Ready Models

We define TRMs, in the context of MBT, as models, which can come in diverse forms (e.g., UML) capturing information that is necessary and sufficient to enable MBT, for the purpose of generating test cases and other required information such as test stimuli and expected test results. TRMs are test ready in the sense that they are amenable for test generation. TRMs are models in the sense that they capture required information at an abstract level. It is, therefore, important to differentiate TRMs from models specifying test cases themselves.

TRMs are often particularly designed for the purpose of testing by test engineers, rather than system engineers. To be able to develop TRMs, test engineers need to have a certain level of understanding of system behaviors and interactions with their operating environments. In addition, as a modeler, a test engineer needs to establish required modeling skills to be able to use proposed modeling notations (e.g., UML) and a corresponding tool implementation.

Belief Test Ready Models (BTRMs) are a specific type of TRMs with uncertainty information explicitly captured as part of the models. Such uncertainty information is specified by test engineers to capture their beliefs about model elements of a BTRM, associated with one or more uncertainties due to "lack of knowledge" about a BTRM when the BTRM was developed. *Belief* and *Uncertainty* should be measured/quantified such that an MBT technique can benefit from the quantified belief and uncertainty information by proposing uncertainty-wise test generation and optimization techniques. In order to do so, a modeling methodology particularly designed for specifying and quantifying uncertainty is needed. Such a methodology ideally should be easy to use, sufficiently expressive in terms of capturing all required information, and ideally based on standard modeling notations, as we described earlier. Most importantly, *uncertainty modeling* is the key to developing BTRMs and for enabling *UWTG* required to capture uncertainty information explicitly.

2.2.1.2 Uncertainty Modeling in General

Uncertainty Modeling in software engineering is a new modeling paradigm; therefore, new modeling notations, tools, and methodologies are required to enable the specification and modeling of uncertainty and its related concepts for the purpose of analyzing any uncertainty-related aspects of a system, and generating other artifacts (e.g., test cases) by taking the uncertainty information into account. Such an uncertainty-modeling paradigm ideally should support the specification/modeling of uncertainty at different levels of abstractions, depending on particular needs. For example, at the requirements engineering phase, uncertainty should be explicitly captured as an important attribute of requirements, similar to their other attributes such as *Verifiability* and *Completeness*. This is however not the current practice as the community lacks understanding of the importance of explicitly specifying not only requirements but also associated uncertainties and confidence of requirement engineers for the specified requirements. Another example is that uncertainty can be specified as part of TRMs, which are collectively named as BTRMs as we discussed earlier.

To understand what uncertainty is in the context of software engineering in general, recently, a conceptual model, named as U-Model [20], has been proposed to define uncertainty, belief, belief agent, indeterminacy source, and other related concepts. U-Model was proposed from the perspective of software engineers (e.g., requirements engineers, test engineers) to understand uncertainty. In other words, U-Model was defined from a *subjective* perspective. U-Model can be considered as the first attempt toward the direction of pursuing a common understanding of uncertainty in the software engineering community. There are a few possibilities of applying U-Model. First, it can be extended for particular purposes. For example, U-Model can be extended for understanding and classifying uncertainty in CPSs. Second, U-Model can be implemented as an independent domain-specific language. Third, U-Model can be integrated with other modeling notations. For instance, a UML uncertainty profile, named as UUP, has been developed, based on U-Model, to facilitate the development of BTRMs, as discussed in Ref. [21]. Another example is to integrate U-Model with other purpose-specific modeling/specification methodologies, such as use case modeling. In Refs. [22,23], the authors presented a framework, named as U-RUCM, for the purpose of capturing uncertainties in use case models and facilitating the discovery of unknown uncertainties. It is worth mentioning that U-RUCM was built on a well-established use case modeling paradigm (Zen-RUCM) [24–26]. In that case, U-Model was used as the basis to develop, for instance, uncertain alternative flows and uncertain sentences. Among all of them, the key motivation of U-Model was to facilitate the common understanding of uncertainty in the software engineering community. As long as the common understanding is built, we believe uncertainty modeling will form a new modeling paradigm. Consequently, tools and corresponding standards will be developed and proposed such that we can put uncertainty modeling into practice.

Considering the fact that SysML has been widely used for system modeling [27,28], in the context of uncertainty-wise testing of CPSs, it is therefore also important to seek opportunities of bringing uncertainty modeling to SysML. The latest version of SysML 1.4 includes a very limited capability of modeling distribution (via, e.g., "Normal," "Uniform"), which is, however, from the perspective of being "complete," and when comparing with U-Model at the conceptual level, there is a huge potential to extend this capability by (1) introducing more comprehensive list of probability distribution types (e.g., triangular distribution) and other types of measures (e.g., fuzziness and ambiguity) and (2) introducing other U-Model concepts (e.g., belief, indeterminacy source) to SysML. Another possibility is to extend the requirements modeling part of SysML by introducing subjective uncertainty to textual requirement statements. The idea is similar to U-RUCM, which however particularly focuses on use case modeling, not only on textual requirement statements.

In summary, we believe U-Model can be a starting point for developing a standardized uncertainty-modeling paradigm such that it can be integrated with other modeling solutions for different purposes in various contexts. OMG has started the standardization activities of Uncertainty Modeling [29].

2.2.1.3 Modeling Belief Test Ready Models

As we discussed earlier, BTRMs are the key input to enable MBT of a system under uncertainty. Therefore, practically useful and meaningful modeling notations, tools, and methodologies are required to enable the development of such BTRMs. In Ref. [21], the authors have developed a UML-based, uncertainty-wise modeling framework, named as *UncerTum*. *UncerTum* was proposed to enable the development of BTRMs for the purpose of testing CPSs under uncertainty. In other words, it is not a generic modeling solution for developing general BTRMs. However, some aspects of *UncerTum* can serve for general BTRM modeling. Nevertheless, we are not aware of any other uncertainty-modeling solutions for developing BTRMs.

UncerTum is built on a set of well-known modeling technologies, including UML, MARTE, OCL, and UTP. More specifically, in the current implementation of *UncerTum*, UML class diagrams and state machines are the artifacts, on which uncertainty information is attached. The rationale behind selecting these two particular UML notations is because they are commonly used for enabling MBT [15,30–32]. The core of *UncerTum* is the UML Uncertainty Profile (UUP), as we discussed earlier, which is built on U-Model. UUP enables the modeling of belief, uncertainty, and measurement, and it also defines an extensive list of model libraries for specifying uncertainty patterns (e.g., *Periodic*), measures (e.g., *Probability*), and time (borrowed from MARTE). *UncerTum* has been integrated with an MBT framework named as *UncerTest* [33], which will be discussed in details in Section 2.2.2. One can adopt, adapt, or extend *UncerTum* for other UML modeling notations such as UML sequence and activity diagrams in case such notations are needed to support particular MBT techniques such as Refs. [34,35]. One can also adopt, adapt, or extend *UncerTum* for developing

BTRMs for testing other types of systems, in addition to CPSs. Other well-applied modeling notations (e.g., SysML) can be also extended to enable uncertainty modeling; therefore, *UncerTum* is a very nice example to exemplify how such an extension can be developed.

2.2.1.4 Modeling Uncertainty in a Life Cycle

Conforming to classical software/system development life cycles and following the transformation principles of MDE, there is a possibility to develop a full uncertainty-modeling/tooling/methodology chain. Such a chain of uncertainty modeling can start from specifying uncertainty requirements, to modeling uncertainty at the architecture and design level, and all the way for developing BTRMs. Developing such a toolchain is very useful. For example, at the requirements engineering phase, uncertainty requirements are specified by requirements engineers, who explicitly indicate, by using certain uncertainty requirements specification methodology (e.g., U-RUCM), "places" where they lack confidence about a requirement statement and their confidence level, indicating to what extent they lack knowledge on what they are specifying. The ultimate objective is to eventually eliminate explicitly specified uncertainties at the requirements engineering phase as much as possible, which is however in practice not always possible. Instead, one might want to discover as many uncertainty requirements as possible and as early as possible, such that mitigation plans can be made before it gets too late. Therefore, automated uncertainty requirements analyses are preferred in such context, as briefly discussed in Ref. [22]. For uncertainty requirements that have to be carried on to the next phase of the development life cycle, software/system designers and developers then need to keep in mind such uncertainty requirements and find ways to mitigate them if possible. During the testing phase, uncertainty requirements can be transferred as part of BTRMs such that systems can be tested against those uncertainty requirements.

2.2.1.5 Verifying and Evolving Belief Test Ready Models

TRMs are expected to be correct and complete in terms of enabling MBT before they are used as the input of an MBT solution. Therefore, it is necessary and important to verify their correctness and completeness, both syntactically and semantically. Nowadays, most of the existing modeling tools can facilitate automated checking of syntactic correctness of models. However, dedicated techniques and tools are needed to check their semantics and their completeness. Especially when verifying BTRMs, correctness and completeness of the uncertainty information of a BTRM have to be verified. Therefore, novel techniques for verifying uncertainty aspects of BTRMs or other belief models (e.g., uncertainty requirements specified in U–RUCM) are urgently needed. However, currently, we are not aware of such solutions and future development in this research direction is welcomed.

As we discussed earlier, a BTRM model captures *subjective* uncertainty at the beginning as they were specified from the perspective of one or more modelers, who are defined as belief agents in U–Model. However, there is a possibility to continuously evolve BTRMs when more and more evidence is collected to update belief agents' understanding of uncertainty, which eventually and ideally should be reflected in future revisions of original BTRMs. There might be many ways to achieve this objective. Here, in this section, to inspire readers, we introduce one of such methodologies, which is named as *UncerTolve* [36,37]. *UncerTolve* was developed in the context of supporting MBT of CPSs under uncertainty, with the objective of evolving BTRMs continuously, motivated by the hypothesis that high-quality BTRMs have higher chance to result in high-quality test cases. In Refs. [36,37], the authors proposed a framework for interactively evolving BTRMs, specified with *UncerTum* (Section 2.2.1). An original BTRM means a BTRM developed by a test engineer with her/his beliefs containing *subjective* uncertainties captured as part of the model. *UncerTolve* aims to mine *objective* uncertainties from real data of the system, which can be collected in various ways, including obtaining real operational data from previous deployments of the same system and/or the same types of systems and obtaining test logs from previously executed test cases. Such data that reflect execution results of the systems are objective and can be used for deriving meaningful information that can be used to improve the quality of the BTRM. For example, previously unknown uncertainties can be discovered from such data; the original BTRM can be refactored with more precisely defined state invariants (specified as OCL constraints, if the BTRM was developed with *UncerTum*, i.e., extended UML state machines); and the uncertainty measurements of the original BTRM can be improved by reflecting not just subjective aspects but also objective aspects. *UncerTolve* relies on machine learning techniques to achieve these objectives and we believe there exist many other techniques for enabling the evolutions of BTRMs. Future investigations are definitely needed toward this research direction.

2.2.2 Uncertainty-Wise Test Generation

In this section, we discuss *UncerTest* and our visions beyond *UncerTest*, including interesting future research directions.

2.2.2.1 UncerTest and Beyond

As we discussed earlier, test generation includes two parts: test case generation and test data generation. In the context of MBT, TRMs are used as the basis to generate both test data and test cases. Various strategies have been proposed in the literature. In the following section, we discuss potentials of using uncertainty information to guide the generation of test cases and test data.

Typical test generation in the context of MBT with TRMs specified mainly as state machines rely on structural coverage criteria such as state coverage, transition coverage, and path coverage [11,12,38,39], which are defined on TRMs. For example, in *UncerTest*, BTRMs are queried to generate abstract test cases from BTRMs by following two commonly used test case generation strategies: All Simple Paths (ASP) and All Paths with Maximum Length (APML). Of course, *UncerTest* can be integrated with other generation strategies and further investigation is required to understand which strategies are better in which situations.

In the context of *UWTG*, there are in principle two kinds of mechanisms to generate abstract test cases. One mechanism is about generating abstract test cases from BTRMs as other MBT techniques do, but during the generation, each test case is attached with uncertainty information (e.g., the number of uncertainties covered by the test case). One of the challenges of MBT is that an MBT technique often generates a large number of test cases; therefore, finding a way to reduce the number of test cases to be eventually executed is very critical. Hence, uncertainty information attached to each test case can be used to define test optimization heuristics/strategies, as we will discuss in Section 3. In other words, this test case generation strategy itself is not specific to *UWTG*, but deriving/calculating uncertainty information for test cases by taking uncertainty information specified in BTRMs as the input is uncertainty-wise. It is also important to mention that different theories (e.g., probability theory [40] and uncertainty theory [41]) can be applied during the uncertainty information generation process for test cases. Which theory to apply depends on how the uncertainty information was obtained and specified as part of BTRMs at the first place, which determines whether prerequisites of a specific theory are satisfied. For example, applying probability theory requires that sufficient data observed from previous executions of the System Under Test (SUT) such that uncertainty can be measured with probability. Otherwise, one might consider using uncertainty theory [41].

The second mechanism utilizes uncertainty information specified as part of BTRMs to guide the generation of the test case. In other words, one needs to propose uncertainty-wise test generation strategies. Such a strategy can be designed, for example, to find a minimum number of paths but covering as many uncertainties as possible, with the ultimate objective of minimizing the number of test cases to be executed eventually. Notice that the difference with the first mechanism is that test optimization (e.g., test selection) is integrated as part of test case generation. The second mechanism defines two sequential steps: generation first and then optimization.

It is hard to say, in a general context, which mechanism is better at the moment, as we, the community, need to collect more experience about *UWTG*, particularly about diverse application contexts of *UWTG*, uncertainty-wise test generation, optimization strategies, and accompanied empirical studies.

Test data generation [5,42] is also an important part of test generation. For generating executable test cases, new test data (often named as artificial data) can be generated by following some strategies or actual data that has been taken from previous operations. Sometimes, it is also possible to combine artificial data and actual data. Cost-effectively generating effective test data is always a challenge in testing. Therefore, in the literature, many test data strategies have been proposed, including random test data generations [42], boundary-value analysis-based test data generations [42], and equivalent partitioning-based test data generations [43]. Search algorithms have been also successfully used in the past for test data generation, as reported in Refs. [7,8,44]. There also exist constraint solvers that help to generate valid test data. One of such constraint solver is EsOCL [17], which was built based on search algorithms to efficiently find valid test data. In the current implementation of *UncerTest*, a random test data generation strategy was applied to generate test data for generated abstract test cases. However, in the future, it is worth investigating uncertainty-wise test data generation strategies such that high quality and uncertainty-oriented test data can be generated. For example, uncertainty-related results of test executions (e.g., the number of observed uncertainties) can be used for future test data generation.

2.3 Tool Support

Automation is all about tool support. Therefore, it is important to recognize the importance of tooling in the context of test generation. There exist a large number of test generation tools. Well-known open source test generation tools include Refs. [19,45–47] and commercial test generation tools include Refs. [48–50].

Tooling is especially important for MBT, as applying modeling tools to produce TRMs is not always straightforward. Therefore, the usability, applicability, and expressiveness of a modeling tool should be taken into account when producing an MBT framework. From industrial practitioners' perspective, it is important to select a tool with good applicability, usability, and expressiveness for easier adoption. The success of any automated MBT solution depends on these properties of its tool implementation, based on which a methodology is often provided to further ease the process of applying an MBT solution, especially considering such a solution is often nontrivial.

2.4 Standardization

Considering the fact that standards play an increasingly important role in industrial practices, we conducted a survey to understand whether there are standardization bodies and standards that are relevant for uncertainty modeling. Motivating by this, we identify standardization bodies such as European Telecommunications Standards Institute (ETSI) and Object Management Group (OMG) and standards that are relevant to modeling and testing CPSs under uncertainty. In this section, we first provide an overview of relevant standards and standardization bodies. Second, we summarize the procedure of selecting standardization bodies. Finally, we present a list of the selected standardization bodies in the Appendix.

2.4.1 Context

Standardization bodies are commonly classified, according to their geographical designation, into three types: international, regional, and national standardization bodies [51]. International standardization bodies develop international standards. There are four most well-known and well-established international standardization bodies: the International Organization for Standardization (ISO), the International Electrotechnical Commission (IEC), the International Telecommunication Union (ITU), and the IEEE Standards Association. Under these four standardization bodies, a large number of standards have been defined. For the regional-level standardization bodies, we only consider EU standards, among which the ETSI produces a lot of standards in ICT. We do not include any national standardization body into the consideration, as standards produced by national-level standardization bodies inherently have limited application scopes, in comparison to international and EU standards.

We aim to define methodologies to test CPSs under uncertainty. One important mean to achieve this objective is to rely on MBE technologies. Therefore, the first criterion of selecting relevant standardization bodies such as the OMG is to include standards in MBE field (C1). As we aim to devise testing methodologies, the second selection criterion is to include standardization bodies and standards that are relevant to the testing field (C2). For example, one of the highly relevant standards is the OMG's UML Testing Profile (UTP) 2 [52]. The third selection criterion is to select standardization bodies and standards that are relevant to software-intensive systems, particularly CPSs (C3).

2.4.2 Relevant Standardization Bodies and Standards

Based on the selection criteria C1–C3 presented earlier, we started from screening through the standards of the standardization bodies: ISO, IEEE, IEC, JCGM, OMG, ETSI, and OASIS from two perspectives: modeling uncertainty and CPS, and testing uncertainty and CPS. As the results of the first step, we preselected a set of standardization bodies and standards as shown in the first column of the Appendix. In the second column of the Appendix, we indicate whether a standard is for modeling, testing, Model-Based Testing (MBT), or others. In the fourth column of the Appendix, we indicate whether a specific standard explicitly defines or describes Uncertainty (including Probability) and Uncertainty Measurement.

As one can see from the Appendix, in terms of modeling, OMG defines standards on system and software modeling: UML [53], SysML [54], MARTE [55], OCL [56], and MOF [57]. ISO/IEC defines UML [58], OCL [59], and KDM [60] modeling notations, which are also defined and maintained by OMG. In addition, ISO/IEC also defines RM-ODP [61] for enabling conceptual modeling of complex systems such as CPSs.

In terms of testing and MBT, OMG defines UTP. ETSI also defines standards on model-based testing: ETSI TR 102 840 V1.2.1 [5], ETSI ES 202 951 V1.1.1 (2011-07) [62], and ETSI EG 201 015 V2.1.1 (2012-02) [63] as shown in the Appendix. ISO/IEC/IEEE 29119 [64] is a widely recognized standard for testing. In addition, ISO/IEC joined the effort to define the ISO/IEC 9646 series [65] for supporting conformance testing of OSI. As shown in the Appendix, a number of standards have been also defined in IEEE from the aspects of system and software verification and validation [66], test documentation [67], unit testing [68], and classification of Software Anomalies [69].

In the Appendix, we also include standards (from ISO, IEC, and/or IEEE under Other) that are relevant to various aspects of system and software engineering, including vocabulary, architecture description, development life cycle, risk management and assessment, and quality assurance. Particularly, we collected standards that are relevant to Uncertainty and Uncertainty Measurement, as indicated in the column "Uncertainty" of the table. ISO 61508 [70], OMG SysML [54], and MARTE [55] define concept Probability, which is one type of uncertainty measures. OMG SACM defines Evidence, Confidence, and Confidence Level, which are all relevant to uncertainty and several concepts defined in U-Model. ISO/IEC and JCGM defined few standards on Uncertainty Measurement. The concept of Uncertainty and few relevant concepts are explicitly defined in ISO 31000 (as shown in the Appendix).

2.4.3 Conclusion

Based on the results of the above-presented survey, we can conclude that there does not exist any standard, which can be used as it is for modeling uncertainty and relevant concepts both subjectively and objectively. We, therefore, initiated the standardization activity of Uncertainty Modeling at OMG. Details can be found in Ref. [29].

3. UNCERTAINTY-WISE MULTIOBJECTIVE TEST OPTIMIZATION

In this section, we detail uncertainty-wise multiobjective test optimization from the following aspects. Section 3.1 provides a formal definition for multiobjective search problem followed by formally defining uncertainty-wise multiobjective search problem (Section 3.2). Section 3.3 presents uncertainty-wise multiobjective test optimization in detail, while Sections 3.4 and 3.5 present uncertainty-wise multiobjective test set minimization and test case prioritization, respectively. Section 3.6 illustrates the above-mentioned uncertainty-wise multiobjective test set minimization and test case prioritization with examples, and Section 3.7 provides a list of examples for cost–effectiveness measures based on the state of the art. Last, Section 3.8 presents a set of guidelines when applying search-based techniques for addressing multiobjective test optimization problems, which is also applicable for uncertainty-wise multiobjective test optimization.

3.1 Multiobjective Search Problem

Generically speaking, a multiobjective optimization problem has a set of objectives (Obj) to meet that often have trade-off relationships among them.

$Obj = \{o_1, o_2, \ldots, o_{non}\}$, where non is the total number of objectives for a particular optimization problem. Each objective o_i is measured with a measure classified into one of the following three categories: (1) $Cost$ Measures, (2) $Effectiveness$ Measures, and (3) $Efficiency$ Measures. Our aim is always to decrease the cost, increase the effectiveness, and/or increase the efficiency. Mathematically speaking, we have a set of $Cost$ measures and $Effectiveness$ Measures:

$CostMeasure = \{cm_1, cm_2, \ldots, cm_{ncm}\}$, where ncm is the total number of cost measures.

$EffectMeasure = \{em_1, em_2, \ldots, em_{nem}\}$, where nem is the total number of effectiveness measures.

In addition, a set of efficiency measures can also be defined:

$EfficiencyMeasure = \{ec_1, ec_2, \ldots, ec_{nec}\}$, where nec is the total number of efficiency measures. An efficiency measure is typically calculated as $Effectiveness$ per $Cost$. Assuming that all combinations of effectiveness measures ($EffectMeasure$) per cost ($CostMeasure$) are valid, then $nec = ncm * nem$. However, not all such combinations are always valid and thus $nec < ncm * nem$.

Assuming that there are a set of possible optimization solutions:

$PS = \{ps_1, ps_2, \ldots, ps_{nps}\}$, where nps is the total number of valid solutions that are typically huge and require an optimized way to the find the best solution. Let us suppose we have three functions: (1) $CMF()$ is a function that takes input a solution ps_i from PS and an objective measured with cost measure cm_i and returns an overall cost objective value for the solution measured with cm_i, (2) $EMF()$ is a function that takes input a solution ps_i from PS and an objective measured with an effectiveness measure em_i and returns the overall effectiveness objective value for the solution measured using em_i, and (3) $ECF()$ is a function that takes input a solution ps_i from PS and an objective measured with an efficiency measure ec_i and returns the overall efficiency value of the solution measured using ec_i.

A multiobjective search problem then can be formally defined as finding the best solution ps_k out of PS such that it holds the following three conditions:

Condition 1: $\forall_{cm}$ in CostMeasure, $\forall_{ps}$ in PS—ps_k, CMF(ps_k, cm) $\leq$ CMF (ps, cm). Notice that $\forall_{ps}$ in PS refers to the explored solutions, of all the nps number of solutions.

Condition 2: $\forall_{em}$ in EffectMeasure, $\forall_{ps}$ in PS—ps_k, EMF(ps_k, em) $\geq$ EMF (ps, em). Once again, $\forall_{ps}$ in PS refers to the explored solutions, of all the nps number of solutions.

Condition 3: $\forall_{ec}$ in EfficiencyMeasure, $\forall_{ps}$ in PS—ps_k, ECF(ps_k, ec) $\geq$ ECF(ps, ec). Once again, $\forall_{ps}$ in PS refers to the explored solutions, of all the nps number of solutions.

For instance, one of our previous works [71] employed multiobjective search (e.g., NSGA-II [72]) for investigating the resource-ware test case prioritization problem by taking into test resource (e.g., hardware) usage into account. To address such a problem, we proposed and formally defined three effectiveness objectives (i.e., test resource usage, fault detection capability, and prioritization density) and one cost measure (i.e., total time). The fitness function was further defined by considering these four cost-effectiveness measures, which was further incorporated into seven multiobjective search algorithms (e.g., NSGA-II) for empirical evaluation in terms of their performance and scalability.

3.2 Uncertainty-Wise Multiobjective Search Problem

An uncertainty-wise multiobjective search problem is a specialization of the generic multiobjective search problem, where an additional dimension of objectives is added to the search problem, i.e., Uncertainty. Assuming, we have a set of Uncertainty Measures:

UncerMeasure = $\{um_1, um_2, ..., um_{num}\}$, where num is the total number of measures that can be used to measure the uncertainty of a solution from different uncertainty perspectives. Assume that there is a function UMF() that takes as input a solution ps_i from PS and an objective measured with an uncertainty measure um_i and returns the overall uncertainty value associated with the solution measured using um_i.

An uncertainty-wise multiobjective search problem has to meet an additional condition described below:

Condition 4: $\forall_{um}$ in UncerMeasure, $\forall_{ps}$ in PS—ps_k, UMF(ps_k, um) $\leq$ or $\geq$ UMF(ps, um). Once again, $\forall_{ps}$ in PS refers to the explored solutions, of all the nps number of solutions.

3.3 Uncertainty-Wise Multiobjective Test Optimization

Building on the previous definitions, an uncertainty-wise multiobjective test optimization aims at optimizing a set of test cases $TC = \{tc_1, tc_2, ..., tc_{ntc}\}$, where ntc is the total number of test cases to optimize based on cost, effectiveness, efficiency, and uncertainty objectives.

3.3.1 Cost, Effectiveness, Efficiency, and Uncertainty Attributes

Test cases have a set of *Cost*, *Effectiveness*, *Efficiency*, and *Uncertainty* attributes associated with them:

CostAtt $= \{cat_1, cat_2, ..., cat_{ncat}\}$, where *ncat* is the total number of cost-related attributes, and each cat_i has exactly one type that can be measured with a basic data type such as *Integer*, *Real*, and *Boolean* or an advanced data type.

EffectAtt $= \{efat_1, efat_2, ..., efat_{nefat}\}$, where *nefat* is the total number of effectiveness-related attributes, and each $efat_i$ has exactly one type that can be measured with a basic data type such as *Integer*, *Real*, and *Boolean* or an advanced data type.

EfficiencyAtt $= \{ecat_1, ecat_2, ..., ecat_{necat}\}$, where *necat* is the total number of efficiency-related attributes and each $ecat_i$ has exactly one type that can be measured with a basic data type such as *Integer*, *Real*, and *Boolean* or an advanced data type.

UncertaintyAtt $= \{uat_1, uat_2, ..., uat_{nuat}\}$, where *nuat* is the total number of uncertainty attributes and each uat_i has exactly one type that can be measured with a basic data type such as *Integer*, *Real*, and *Boolean* or an advanced data type.

3.3.2 Cost, Effectiveness, Efficiency, and Uncertainty Attribute Values

Each test case tc_i in TC has four sets of associated values that can be used to calculate the four types of objectives, i.e.:

CostVal$_i = \{cvali_1, cvali_2, ..., cvali_{ncat}\}$, where *ncat* is the total number of cost-related attributes measured with the *Cost* measures.

EffectVal$_i = \{efvali_1, efvali_2, ..., efvali_{nefat}\}$, where *nefat* is the total number of effectiveness-related attributes measured with the *Effectiveness* measures.

EfficiencyVal$_i = \{ecvali_1, ecvali_2, ..., ecvali_{necat}\}$, where *necat* is the total number of efficiency-related attributes measured with the *Efficiency* measures.

$UncertaintyVal_i = \{umvali_1, umvali_2, ..., umvali_{nuat}\}$, where $nuat$ is the total number of uncertainty-related attributes measured with the $Uncertainty$ measures.

3.3.3 Calculation of Cost, Effectiveness, Efficiency, and Uncertainty Objectives

In the context of a test optimization problem, the input for the search is the set of test cases, i.e., TC and its associated sets of cost, effectiveness, efficiency, and uncertainty attribute values for the ntc number of test cases. A solution ps_k in PS, in this context, would be a set of test cases from TC, i.e., either a subset (if it is a test set minimization problem) or the same subset but with the best order to execute the same set of test cases (prioritization problem).

The calculation of an objective o_i (Cost/Effectiveness/Efficiency/Uncertainty) can be performed in different ways and is often dependent on the problem being solved. A simplest and straightforward way to calculate an objective is to take an average of the solution. For example, a cost measure cm_i corresponding to a cost attribute, e.g., cat_k can be calculated as follows:

$cm_i = (\sum_{j=1 \text{ to } ntc} cvalj_k)/ntc$, where $cvalj_k$ represents a value for a jth test case in TC corresponding to the cat_k cost attribute.

Similarly, for an effectiveness measure em_i corresponding to an effectiveness attribute, e.g., $efat_k$ can be calculated as follows:

$em_i = (\sum_{j=1 \text{ to } ntc} efvalj_k)/ntc$, where $efvalj_k$ represents a value for a jth test case in TC corresponding to the $efat_k$ effectiveness attribute.

For an efficiency measure ec_i corresponding to an efficiency attribute, e.g., $ecat_k$ can be calculated as follows:

$ec_i = (\sum_{j=1 \text{ to } ntc} ecvalj_k)/ntc$, where $ecvalj_k$ represents a value for a jth test case in TC corresponding to the $ecat_k$ effectiveness attribute.

For an uncertainty measure um_i corresponding to an uncertainty attribute, e.g., uat_k can be calculated as follows:

$um_i = (\sum_{j=1 \text{ to } ntc} umvalj_k)/ntc$, where $umvalj_k$ represents a value for a j_{th} test case in TC corresponding to the uat_k uncertainty attribute.

Given that each objective, i.e., cm_i, em_i, ec_i, and um_i, may take a value in different ranges, a common practice is to normalize the values calculated for the objectives. There are two commonly used normalization functions in the context of search-based multiobjective for software engineering problems depending on two conditions:

Condition 1: If maximum and minimum values for a calculated objective are known, then the following normalization function must be used:

$N1(x) = (x - x_{min})/(x_{max} - x_{min})$, where x_{max} represents the known maximum value that can be taken by the variable x and x_{min} represents the known minimum value that can be taken by the variable x.

Condition 2: If the maximum and minimum values for a calculated objective are not known, then one of the following normalization functions should be used:

$N2(x) = (x/(x + \beta))$, where β is any value greater than 0 [73,74].

Notice that there exists another normalization function for this condition: $N3(x) = 1 - \alpha^{-x}$, where $\alpha = 1.001$ is typically used [73,74]. However, the experiments reported in Refs. [73–75] suggest using *N2* instead of the *N1* normalization function.

Our ultimate aim is to minimize cost and maximize effectiveness and efficiency, whereas at the same time minimize (or maximize) uncertainty depending on the problem at hand, i.e.,

$\forall_{cm}$ in *CostMeasure*, Minimize(cm) $\wedge$ $\forall_{em}$ in *EffectMeasure*, Maximize(em) $\wedge$ $\forall_{ec}$ in *EfficiencyMeasure*, Maximize(ec) $\wedge$ $\forall_{um}$ in *UncerMeasure*, Minimize/Maximize(um).

3.4 Uncertainty-Wise Multiobjective Test Set Minimization

In the case of *UWTM*, our aim is to find a minimum number of test cases (*mtc*) out of the total number of test cases *ntc* in *TC* that meet all the required cost, effectiveness, efficiency, and uncertainty objectives. Thus, the number of test cases (*mtc*) in a solution in *PS* can be any combination of a number of test cases in *TC*, i.e., from 1 to $ntc - 1$. In this way, the possible number of minimization solutions can be calculated as follows:

$$nps = {}^{ntc}C_1 + {}^{ntc}C_2 + \cdots + {}^{ntc}C_{ntc-1} = 2^{ntc} - 1$$
$$PS = \{ps_1, ps_2, \ldots, ps_{2ntc-1}\}$$

If $mtc = ntc$, then it means that there is no minimization at all. Irrespective of which cost, effectiveness, efficiency, and uncertainty measures are selected for *UWTM*, a mandatory cost objective must be defined that considers test set minimization. Depending on the problem at hand, such an objective can be defined in different ways. One way to define such objective is to calculate the percentage of test set minimization (TMP), which can be calculated as:

TMP $= 1 - ntc_i/ntc$, where ntc_i is the total number of test cases in the ps_i solution in PS and ntc is the total number of test cases in TC.

Another way of defining a cost objective measuring test set minimization can simply be a number of minimized test cases (NMTC):

NMTC $= ntc_i$ is the number of minimized test cases in a solution ps_i.

Given the fact that TMP is a percentage, it will produce a value between 0 and 1 and does not need normalization. In the case of NMTC, a normalization function may be needed to scale its value between 0 and 1. A common problem with both TMP and NMTC is that a search algorithm will favor a solution with the lower number of test cases and can potentially lead to selecting no test cases at all. In order to avoid search reaching to such a situation, various mechanisms can be employed. For example, a typical way is to control a search algorithm to produce solutions with at least x number of test cases, where x is always greater than 1.

An uncertainty-wise multiobjective test set minimization can formally be defined as finding the best solution ps_k from PS which has a minimum number of test cases that meet the four conditions specialized as follows:

Condition 1: $\forall_{cm}$ in *CostMeasure*, $\forall_{ps}$ in PS—ps_k, CMF(ps_k, cm) $\leq$ CMF(ps, cm) and $mps_k < mps$ and $mps_k \neq 0$, where mps_k is the number of test cases in the ps_k solution and mps is the number of test cases in the ps solution. Notice that $\forall_{ps}$ in PS refer to the explored solutions of all the nps number of solutions.

Condition 2: $\forall_{em}$ in *EffectMeasure*, $\forall_{ps}$ in PS—ps_k, EMF(ps_k, em) $\geq$ EMF(ps, em) and $mps_k < mps$ and $mps_k \neq 0$, where mps_k is the number of test cases in the ps_k solution and mps is the number of test cases in the ps solution. Once again, $\forall_{ps}$ in PS refers to the explored solutions of all the nps number of solutions.

Condition 3: $\forall_{em}$ in *EfficiencyMeasure*, $\forall_{ps}$ in PS—ps_k, ECF(ps_k, ec) $\geq$ ECF(ps, ec) and $mps_k < mps$ and $mps_k \neq 0$, where mps_k is the number of test cases in the ps_k solution and mps is the number of test cases in the ps solution. Once again, $\forall_{ps}$ in PS refers to the explored solutions, of all the nps number of solutions.

Condition 4: $\forall_{um}$ in *UncerMeasure*, $\forall_{ps}$ in PS—ps_k, UMF(ps_k, um) $\leq$ or $\geq$ UMF(ps, um) and $mps_k < mps$ and $mps_k \neq 0$, where mps_k is the number of test cases in the ps_k solution and mps in the number of test cases is the ps solution. Once again, $\forall_{ps}$ in PS refers to the explored solutions, of all the nps number of solutions.

3.5 Uncertainty-Wise Multiobjective Test Case Prioritization

Uncertainty-Wise Multiobjective Test Case Prioritization (UWTP) is concerned with prioritizing the test cases in a specific order to execute. This is in contrast with *UWTM*, where we wanted to minimize the number of test cases without considering any order of execution. However, test set minimization and test case prioritization may be combined together. There are three possible ways: (1) minimize test cases and then prioritize the minimized test cases, (2) prioritize test cases and then minimize them, and (3) prioritize and minimize the test cases at the same time.

Generally speaking, when performing *UWTP*, our aim is to find a best order to execute the total number of test cases *ntc* that are available in *TC* such that it meets all the required cost, effectiveness, efficiency, and uncertainty objectives. Thus, as opposed to the test set minimization problem, where the number of test cases (*mtc*) in a solution in *PS* can be any combination of number of test cases in *TC*, *UWTP* keeps the number of test cases to the same as the total number of test cases in *TC*, i.e., *ntc*. Thus, in the case of *UWTP*, each test case can possibly appear at each location, and thus the possible number of prioritization solutions can be calculated as follows:

$$nps = (ntc)^* (ntc - 1)^* (ntc - 2)^* \cdots 1 = ntc!$$
$$PS = \{ps_1, ps_2, \ldots, ps_{ntc!}\}$$

Regardless of the selection of cost, effectiveness, efficiency, and uncertainty measures for an *UWTP*, a mandatory effectiveness objective must be defined that calculates the effectiveness of prioritization. Depending on the problem at hand, such a prioritization objective can be defined in a variety of ways. One example to calculate the effectiveness of prioritization is using Prioritization Impact (*PI*) of a location of an order of a test case, which can be calculated as follows:

$PI_i = (ntc - p + 1)/ntc$, where p represents the pth location in a prioritization solution. In other words, the test cases with low cost, high effectiveness, high efficiency, and low uncertainty must be ordered to execute at the earlier positions. For example, at the first location, i.e., $p = 1$, $PI_i = 1$, whereas at $p = ntc$, $PI_i = 1/ntc$. Such number is already scaled between 0 and 1 and thus does not require normalization. A higher value means a high priority.

Another way of defining such an effectiveness objective measuring prioritization impact can simply be the position, i.e., p, where p ranges from 1 to *ntc* and we aim to order the best test case in terms of cost, effectiveness, efficiency, and uncertainty in the earlier position. The fact that p is a number greater than 1 indicates that normalization of p between 0 and 1 may be required. In this particular case, since we know the maximum (i.e., ntc) and minimum (i.e., 1) values for the position, we can use the following normalization function:

$$N1(p) = (p)/(ntc - 1)$$

Typically, when performing test case prioritization, it is not possible to execute all the test cases (i.e., ntc). The execution of the number of test cases is limited by time budget, for example, a certain percentage of test cases. Such time budget can be fixed and encoded in the search problem as an additional constraint that must be satisfied by a solution. Such time budget can be calculated in different ways. For example, as a certain percentage of test cases, e.g., 10% or 20%. Another example is to set this time budget based on the actual execution time, such as in terms of the maximum number of hours.

An uncertainty-wise multiobjective test set prioritization problem can formally be defined as finding the best solution ps_k from PS, which has test cases in the best order to execute such that it meets the following four conditions specialized for $UWTP$:

Condition 1: $\forall_{cm}$ in *CostMeasure*, $\forall_{ps}$ in PS—ps_k, $CMF(ps_k, cm) \leq CMF(ps, cm)$ and $tbg \leq bg$, where tbg represents the overall time budget for the ps_k solution and bg represents the available time budget for test case prioritization. Notice that $\forall_{ps}$ in PS refers to the explored solutions of all the nps number of solutions.

Condition 2: $\forall_{em}$ in *EffectMeasure*, $\forall_{ps}$ in PS—ps_k, $EMF(ps_k, em) \geq EMF(ps, em)$ and $tbg \leq bg$, where tbg represents the overall time budget for the ps_k solution and bg represents the available time budget for test case prioritization. Once again, $\forall_{ps}$ in PS refers to the explored solutions, of all the nps number of solutions.

Condition 3: $\forall_{em}$ in *EfficiencyMeasure*, $\forall_{ps}$ in PS—ps_k, $ECF(ps_k, ec) \geq ECF(ps, ec)$ and $tbg \leq bg$, where tbg represents the overall time budget for the ps_k solution and bg represents the available time budget for the test case prioritization problem. Once again, $\forall_{ps}$ in PS refers to the explored solutions of all the nps number of solutions.

Condition 4: $\forall_{um}$ in *UncerMeasure*, $\forall_{ps}$ in PS—ps_k, $UMF(ps_k, um) \leq$ or $\geq UMF(ps, um)$ and $tbg \leq bg$, where tbg represents the overall time budget for the ps_k solution and bg represents the available time budget for the test case prioritization problem. Once again, $\forall_{ps}$ in PS refers to the explored solutions, of all the nps number of solutions.

3.6 Examples of UWTM and UWTP

In this section, using the formalism defined from Section 3.1 to Section 3.5, we will provide *UWTM* and *UWTP* examples. The both examples will demonstrate the applications of *UWTM* and *UWTP* for testing an industrial CPS case study. The need for these *UWTM* and *UWTP* arose in the context of a project, where we are developing new methods to test CPSs in the presence of uncertainty [76]. Our overall solution was *UWMBT* as defined in Section 2.2. The expected behavior of a CPS under test together with uncertainty was modeled with *UncerTum* (Section 2.2.1), and then using our tool support a set of abstract test cases were generated [33]. Given the complexity of models, the number of generated test cases was large and it was practically impossible to execute all the generated test cases and thus required the development of not only *UWTM* but also *UWTP*. In the following subsections, we will present the *UWTM* and *UWTP* solutions that we developed as examples. Their details and evaluations are reported in Refs. [33,77], and interested readers are suggested to consult these references for further information. The rest of this section is organized as follows: In Section 3.6.1, we present the example of *UWTM* and example of *UWTP* in Section 3.6.2.

3.6.1 Example of UWTM

In our previous work reported in Ref. [33], we defined four *UWTM* problems and demonstrated their application for an industrial CPS case study. Test cases, in this case, were generated from stereotyped UML State machines as discussed in Section 2.2.1 using different uncertainty–wise test generation strategies including ASP with Uncertainty (ASP) and All Paths with Uncertainty and a Fixed Maximum Length (APML) as discussed in Section 2.2.2 and in Ref. [33]. Each generated test case had sets of associated *EffectiveAtt* and *UncertaintyAtt* attributes as listed below:

EffectiveAtt = {*ntran*}, where *ntran* represents the number of transitions covered by a test case.

UncertaintyAtt = {*nun, nuun, um*), where *nun* represents the number of uncertainties in a test case, *nuun* represents the number of unique uncertainties in a test cases, and *um* represents the overall uncertainty of a test case calculated using the uncertainty measure defined in Uncertainty Theory [41].

Notice that it is not always necessary to have all four kinds of attributes (*Cost, Effectiveness, Efficiency,* and *Uncertainty* as discussed in Section 3.3) associated with test cases. In this particular *UWTM*, we defined the following *Cost, Effectiveness,* and *Uncertainty* Objectives:

$Obj = \{PTM, ANU, PUS, AUM, PUU, PTR\}$

CostMeasure $= \{PTM\}$, where *PTM* is the same as *TMP* as defined in Section 3.4 and measures the percentage of minimization.

EffectMeasure $= \{PTR\}$, where *PTR* is an effectiveness measure used to calculate the overall transition coverage achieved by a minimized solution. The formula for calculating *PTR* can be found in Ref. [33].

UncerMeasure $= \{ANU, PUS, AUM, PUU\}$, where *ANU* is an uncertainty measure used to calculate an average number of uncertainties covered by a minimized solution. *PUS* is another uncertainty measure used to calculate the overall uncertainty space covered by a minimized solution. Notice that the concept of uncertainty space is defined in uncertainty theory [41] and is adopted in our work. The formula for calculating *PUS* can be found in Ref. [33]. *AUM* is used to calculate the overall average uncertainty measure of a minimized test set. Recall that uncertainty measure is defined in Uncertainty theory [41] and is adopted in our work. The formula for calculating *AUM* can be found in Ref. [33]. The fourth uncertainty measure we defined is *PUU*, which is used to calculate the overall average number of unique uncertainties covered by a minimized solution. The formula for *PUU* can be consulted in Ref. [33].

Given that not all the uncertainty measures can be used at the same time, we developed four UWTM problems described below:

$UWTM_1 = CostMeasure\ \{PTM\}, EffectMeasure = \{PTR\}$, and *UncerMeasure* $= \{ANU\}$

$UWTM_2 = CostMeasure\ \{PTM\}, EffectMeasure = \{PTR\}$, and *UncerMeasure* $= \{PUS\}$

$UWTM_3 = CostMeasure\ \{PTM\}, EffectMeasure = \{PTR\}$, and *UncerMeasure* $= \{AUM\}$

$UWTM_4 = CostMeasure\ \{PTM\}, EffectMeasure = \{PTR\}$, and *UncerMeasure* $= \{PUU\}$

In the above four *UWTMs*, the cost objectives must be minimized, whereas the effectiveness and uncertainty objectives must be maximized. Depending on the definitions of uncertainty measures, uncertainty objectives may be minimized or maximized. In our context, all the uncertainty objectives must be maximized, e.g., we aim to cover as much uncertainty as possible with a minimized solution.

A variety of multiobjective search algorithms may be used to solve our *UWTM*s. As an initial evaluation, we opted for the most commonly used multiobjective search algorithm, i.e., NSGA-II [72] to solve our *UWTM*s. We used its implementation in the jMetal framework [78]. We empirically evaluated the four UWTMs using an open source case study of SafeHome, where we compared the four *UMTM*s in terms of their effectiveness with mutation testing. Based on the results of our experiment, we found that *UWTM*$_4$ was able to find the best-minimized solution in terms of minimized number of test cases and mutation score (i.e., the number of seeded faults found). Notice that these minimized number of test cases and mutation score are comparable across all the four *UWTM*s. *UWTM*$_4$ managed to achieve *PTM* of 91%, whereas it managed to achieve 100% mutation score [33].

With the best *UWTM*, i.e., *UWTM*$_4$, we minimized the number of test cases for the industrial CPS case study of GeoSports (GS) [79]. We used one use case focusing on testing the implementation of GS in the presence of uncertainties and generated test cases using our *UWMBT* technique with the tool presented in Ref. [33]. With *UWTM*$_4$, we managed to minimize 83.9% of test cases and managed to discover 98 uncertainties when executing the minimized test cases. 18 of these 98 were newly discovered uncertainties. All the details of *UWTM*s and empirical evaluations can be consulted in Ref. [33].

3.6.2 Example of UWTP

Our previous work [77] reported an Uncertainty-wise Test Case Prioritization Problem (*UWTP*) that we defined and demonstrated its application in an industrial CPS case study. The same as the *UWTM*s, test cases were obtained using our *UWMBT* techniques as described in Section 2.2.2 and full details in Ref. [77]. Each generated test case had the following sets of associated *CostAtt*, *EffectiveAtt*, and *UncertaintyAtt* attributes listed below:

CostAtt = {*etime*}, where *etime* is the execution time for a test case

EffectiveAtt = {*ntran*}, where *ntran* represents the number of transitions covered by a test case

UncertaintyAtt = {*oun*, *um*}, where *oun* represents the number of uncertainties observed as the result of execution of the test case and *um* represents the overall uncertainty of a test case calculated using the uncertainty measure defined in Uncertainty Theory [41].

The same as any other multiobjective test optimization problem, it is not mandatory to have all the four types of attributes that we defined. In our *UWTP*, we defined the following *Cost*, *Effectiveness*, and *Uncertainty* objectives:

$Obj = \{PET, ANOU, AUM, PTR\}$

CostMeasure = $\{PET\}$, where *PET* is the overall execution time for the prioritized set of test cases. The formula for calculating *PET* can be found in Ref. [77]. Notice that in this work, we had access to the execution time of each test case; therefore, we opted for time budget based on the overall execution of test cases. For example, we wanted to prioritize test cases based on $x\%$ of the total time budget, where total time budget was equal to the total time to execute all the *ntc* test cases.

EffectMeasure = $\{PTR\}$, where *PTR* is an effectiveness measure used to calculate the overall transition coverage achieved by a minimized solution. The formula for calculating *PTR* can be found in Ref. [77].

UncerMeasure = $\{ANOU, AUM\}$, where *ANOU* is an uncertainty measure used to calculate an average number of observed uncertainties in the last execution of the test cases contained in a prioritized solution. The formula for calculating *ANOU* can be found in Ref. [77]. *AUM* is used to calculate the overall average uncertainty measure of a prioritized test set. Recall that uncertainty measure is defined in Uncertainty theory and is adopted in our work. The formula for calculating *AUM* can be found in Ref. [77].

In terms of *PI*, i.e., Prioritization Index, we used the formula presented in Section 3.5, i.e., $(ntc - p + 1)/ntc$. Our uncertainty-wise multiobjective prioritization problem was represented as follows:

$UWTP = CostMeasure = \{PET\}$, *EffectMeasure* = $\{PTR\}$, and *UncerMeasure* = $\{ANOU, AUM\}$

In the above *UWTP*, the cost objective must be minimized, and the effectiveness objective must be maximized, whereas uncertainty objectives must be maximized as well, i.e., we wanted to prioritize the solutions based on both the modeled subjective uncertainties and the observed uncertainties that were discovered as the result of the last test execution.

In terms of evaluation, we used the same industrial CPS case study, i.e., GeoSports as our *UWTMs*. Once again we selected NSGA-II for solving our *UWTP*, although we also compared the performance of NSGA-II with a greedy algorithm as the comparison baseline [77]. In addition, to assess the scalability of NSGA-II in terms of solving *UWTP* problems with varying complexities, we also simulated, in total, 72 problems ranging from simpler to the complex problems. The results of our evaluation showed that NSGA-II significantly outperformed Greedy for the industrial case study and also for the 72 simulated problems. We concluded that, on average for both industrial and simulated problems, NSGA-II improved prioritization by 22% as compared to the Greedy algorithm. These results showed that NSGA-II is cost effective and scalable for solving our Uncertainty-Wise Test Case Prioritization problem for a variety of problems [77].

3.7 Examples of Cost and Effectiveness Measures

In this section, we present some examples of cost, effectiveness, and uncertainty objectives from the literature, which are related to multiobjective test optimization. Table 2 presents the name and a short description of these objectives together with references where the details can be found.

3.8 Analyzing Results

In this section, we discuss a set of guidelines extracted based on our previous experience of applying Search-Based Software Testing (SBST) to address various multiobjective test optimization problems including uncertainty-wise multiobjective test optimization. We present these guidelines from the following perspectives, which include: (1) *fitness function* that discusses what should be paid attention when defining and formulating fitness function; (2) *multiobjective algorithms* that discuss how to select and compare multiobjective search algorithms; (3) *evaluation metrics* that present how to select appropriate evaluation metrics for evaluating the performance of multiobjective search algorithms; and (4) *statistical tests* that discuss how to select proper statistical tests for analyzing the results obtained by the multiobjective search algorithms in conjunction with the defined fitness function. It is worth mentioning that the summarized guidelines are considered as generic recommendations, which are applicable to any multiobjective test optimization problems (including uncertainty-wise multiobjective test optimization problems).

3.8.1 Fitness Function

It is well recognized that a fitness function is defined for assessing the quality of solutions obtained during the search, and thus defining an appropriate fitness function is of paramount importance to obtain optimal solutions when addressing multiobjective test optimization problems. The first key step to define a proper fitness function is to formally formulate the required

Table 2 List of Cost, Effectiveness, and Uncertainty Objectives

Label	Name	Description
Effectiveness		
E1	The percentage of reduction in test set size (*TMP*)	*TMP* measures the percentage of reduction in test set size compared with original test set [80]. For example, see its use in Refs. [80,81]
E2	The percentage of reduction in fault detection by selected test cases	The number of faults detected by the original test set *TC* is *nf*, whereas the number of faults detected by the optimized test cases is *mf*. Such reduction in fault detection [80] is calculated as $1 - (mf/nf) * 100\%$
E3	The rate of fault detection	"A measure of how quickly a test set detects faults during the testing process" [82]. Such measure is used with test case prioritization
E4	Coverage of statements of code (total coverage and additional coverage)	"A measure of the percentage of statements that have been executed by test cases" [83,84]
E5	Coverage of event (total coverage and additional coverage)	"A measure of the percentage of events that have been executed by test cases" [85,86]
E6	Coverage of call stack	"A test set is represented by a set of unique maximum depth call stacks; its minimized test set is a subset of the original test set whose execution generates the same set of unique maximum depth call stacks" [87]
E7	Operational coverage	"An operational abstraction is a formal mathematical description of program behavior: it is a collection of logical statements that abstract the program's run-time operation. Operational coverage (defined below) measures the difference between an operational abstraction and an oracle or goal specification" [83]

Table 2 List of Cost, Effectiveness, and Uncertainty Objectives—cont'd

Label	Name	Description
E8	*def-use* coverage	"A measure of the percentage of definitions that have been executed by test cases" [88,89]
E9	Coverage of branch (total probability and additional probability)	"A measure of the percentage of branches that have been executed by test cases" [83,89,90]
E10	Coverage of block (total probability and additional probability)	"A measure of the percentage of blocks that have been executed by test cases" [91]
E11	Coverage of functions (total coverage and additional coverage)	"A measure of the percentage of functions that have been executed by test cases" [82]
E12	Coverage of interaction (total probability and additional probability)	"A measure of the percentage of interactions that have been executed by test cases" [85,86]
E13	Percentage of Transition Coverage (PTR)	PRT measures the percentage of the total number of unique transitions covered by the minimized [33] or prioritized [77] subset of test cases
E14	Probability of exposing faults (total probability and additional probability)	FEP [82,92] is the commonly used metric to measure the probability of exposing faults is fault-exposing-potential. It can be calculated as: $FEP = km/tm$, where km is the number of killed mutants by TC, and tm is total mutants in TC.
E15	Probability of faults existence (total probability and additional probability)	"Faults are not equally expected to exist in each function; rather, certain functions are more liable to contain faults than others. This fault proneness can be associated with measurable software attributes. Test cases based on their history of executing fault-prone functions are prioritized taking the advantage of their association" [93]. A metric to measure as a metric of fault proneness is a fault index [93]

Continued

Table 2 List of Cost, Effectiveness, and Uncertainty Objectives—cont'd

Label	Name	Description
E16	Average Percentage of Faults Detected (*APFD*) (only prioritization)	*APFD* [82,94] measures the average percentage of faults detected after prioritization. It can be calculated as: $APFD = 1 - (\sum_{i=1 \text{ to } nf} ps(i))/(nf^* ntc) + 1/(2ntc)$, where *ntc* is number of test cases in *TC*, $F = \{f_1, ..., f_{nf}\}$ is the set of *nf* faults revealed by *TC*, and *ps(i)* is the first position of test cases in the prioritized test cases which reveals fault f_i
E17	Unit-of-fault-severity-detected-per-unit-test-cost *APFDc* (only prioritization)	$APFD_c$ [95] measures the average of percentage of faults detected with cost after prioritization
E18	Customized Test Requirement	There are some characteristics of requirement of test cases, e.g., importance [96], risk [97], customer-assigned priority [96,98,99], implementation complexity (CI) [96,99], requirement changes [99], completeness (CT) [99], traceability (TR) [99], dependency [100], scope [101], property relevance [102]
E19	Feature Pairwise Coverage (*FPC*)	*FPC* is defined to measure how many feature (testing functionalities) pairs can be achieved by a produced solution [81]
E20	Fault Detection Capability (*FDC*)	*FDC* measures the fault detection capability of an obtained solution. In Ref. [81], fault detection is defined as the successful execution rate (manage to detect faults) for a test case in a given time, e.g., a week
E21	Average Execution Frequency (*AEF*)	*AEF* measures the average execution frequency of a solution during a given time (e.g., a week) using the execution frequency of each included test cases [81]
E22	Prioritization Density (*PD*)	*PD* measures how many test cases can be prioritized by a given solution with available test resources (e.g., hardware) [71]

Table 2 List of Cost, Effectiveness, and Uncertainty Objectives—cont'd

Label	Name	Description
E23	Test Resource Usage (*TRU*)	*TRU* is defined to measure how many available test resources (e.g., hardware) can be used by a solution [71]
E24	Mean Priority (*MPR*)	*MPR* measures the importance of the test cases in a solution, which is determined based on the type of test requirements [103]
E25	Mean Probability (*MPO*)	*MPO* measures the likelihood that a solution (including a number of test cases) in terms of detecting faults [103]
E26	Mean Consequence (*MC*)	*MC* measures the impact of a failure of the test cases in a solution that the system can have on the environment [103]
E27	Configuration Coverage (*CC*)	*CC* measures the overall configuration coverage of a solution with a set of a number of test cases [104]
E28	Test API Coverage (*APIC*)	*APIC* measures the overall test API coverage of a solution [104]
E29	Status Coverage (*SC*)	*SC* measures the overall status coverage of a solution [104]
Cost		
C1	Saving Factor, SF (unit is dollars)	This is measured based on "savings in time are associated with savings in dollars through engineer salaries, accelerated business opportunities" [71]
C2	Overall Execution Time (*OET*)	*OET* measures the total execution time for a solution (including a number of test cases) based on the historical execution data [81]
C3	Number of test cases (*NMTC*)	*NMTC* measures the number of test cases that need to be executed. For example, see its use in Ref. [33]
C4	Number of statements	A measure of the number of statements that have been executed by test cases. For example, see its use in Ref. [94]

Continued

Table 2 List of Cost, Effectiveness, and Uncertainty Objectives—cont'd

Label	Name	Description
C5	Total Time (TT)	TT measures the total time cost for a solution including test case execution time and test resource allocation time [71]
C6	Time Difference (TD)	TD measures the difference in time between a pregiven time budget and execution time for a solution (that consists of a number of test cases) [103]
Additional measures		
A1	Average Normalized Number of Uncertainties Covered (ANU)	ANU is used to measure the average normalized number of uncertainties covered by the minimized number of test cases [33]
A2	Percentage of Uncertainty Space Covered (PUS)	PUS measures the percentage of the total set of uncertainty spaces of a belief state machine [21] covered by the minimized subset of test cases [33]
A3	Average Overall Uncertainty Measure (AUM)	AUM measures the average overall uncertainty of the minimized [33] or prioritized [77] subset of test cases
A4	Percentage of Unique Uncertainties Covered (PUU)	PUU measures the percentage of the total number of unique uncertainties covered by the minimized subset of test cases [33]
A5	Average Number of Observed Uncertainties ($ANOU$)	$ANOU$ measures the average number of observed uncertainties by the prioritized test cases [77]

objectives as a set of cost–effectiveness measures based on the specific domain knowledge. For instance, to tackle the multiobjective test set minimization problem [81], we discussed with the test engineers of our industrial partner (with the telecommunication domain) and proposed five objectives that should be taken into account. These five objectives were further formulated as four effectiveness measures (i.e., objective functions) and one cost measure, which includes (1) *test set minimization* to measure the amount of reduction in terms of the number of test sets; (2) *feature pairwise coverage* to measure how many feature (testing functionalities) pairs can be achieved by a solution; (3) *fault detection capability* to measure the capability of detection faults based on the historical data; (4) *average execution frequency* that measures how often the minimized test cases can be executed based on the execution history; and (5) *overall execution time* to measure the overall time taken for executing the minimized test cases.

It is worth mentioning that it is practically possible to further refine the defined cost–effectiveness measures based on particular restrictions from particular domains, and thus it is recommended to first finalize the cost–effectiveness measures (objective functions) before defining fitness function (**Recommendation, R1**). For instance, with respect to the effectiveness measure fault detection capability, it is common in the existing literature to define such measure using the number of detected faults (bugs). However, in the context of our case (i.e., telecommunication domain) [81], such information related to the number of faults is not available from the historical execution data, and thus we proposed to measure the fault detection capability by calculating the number of successful executions out of the total number of executions (a successful execution means faults can be detected by executing a specific test case). The measurement of fault detection capability was further discussed, agreed, and finalized with the test engineers from our industrial partner. Therefore, we recommend practitioners defining the cost–effectiveness measures based on the particular domains since it is sometimes possible that the common measures from the state of the art cannot fulfill the practical requirements from the domains (**R2**).

Once the cost–effectiveness measures are formally formulated, the next step is to incorporate these measures into a fitness function that is used to guide the search toward finding optimal solutions. Based on the state of the art, the fitness function can be defined using two means including (1) directly using the defined cost–effectiveness measures as the fitness function when the Pareto-based multiobjective algorithms (e.g., NSGA-II) are further employed in conjunction with the fitness function; and (2) assigning particular weights to each cost–effectiveness measure and converting the multiobjective test optimization problem to a single-objective test optimization problem if the weight-based multiobjective algorithms are applied (e.g., RWGA). One general guide to recommend is that the first mean should be applied if all the objectives hold equivalent priority, i.e., the objectives are considered as equally important and the second mean is chosen if there are specific priorities (i.e., user preferences) among the objectives in practice, e.g., fault detection capability is more important than the test minimization percentage for the test set minimization problem (**R3**). Notice that the performance of Pareto-based search algorithms will be significantly decreased when the number of cost–effectiveness measures (objectives) is more than six [105] and thus it is suggested to limit the number of objectives less or equal to six if the Pareto-based search algorithms are to be applied (**R4**). However, it is worth mentioning that one of our previous works [106] proposed a new Pareto-based search algorithm (named UPMOA) to incorporate user preferences into NSGA-II, which made it possible for the Pareto-based search algorithms to handle the multiobjective test optimization problems with various user preferences.

In addition, if the weight-based search algorithms are to be employed, there are three weight assignment strategies defined and applied in the existing literature [107], which include: (1) *Fix Weights (FW)* assignment strategy that assigns predefined quantitative weights (between 0 and 1) to each objective (e.g., 0.5 to the fault detection capability) based on the user preferences from domain experts (e.g., test engineers); (2) *Randomly Assigned Weights (RAW)* assignment strategy that randomly generates normalized weights (between 0 and 1) for each test optimization objective that satisfy the defined user preferences; and (3) *Uniformly Distributed Weights (UDW)* assignment strategy that generates normalized weights for objectives, which (1) meet the user preferences and (2) guarantee the uniformity for the generated weights with the aim of ensuring that each search direction is explored with equal chances. The experiment results from Ref. [107] showed that *UDW* can manage to achieve better performance as compared with the other two weight assignment strategies (i.e., *FW* and *RAW*), which can be used as a guide when there is a need to apply these weight assignment strategies for weight-based genetic algorithms (**R5**).

3.8.2 Multiobjective Search Algorithms

After the fitness function is properly defined, the next step is to select the multiobjective search algorithm to incorporate the fitness function before running the search. As mentioned before, Pareto-based and weight-based multisearch algorithms can be used and all these algorithms have intrinsic randomness because of several factors such as initial random population, the use of a crossover, and mutation operators. Thus, it is recommended to repeat each algorithm at least 10 times [73,105], which can ensure that the results obtained by the search algorithms are not achieved at random, thereby reducing the random variations inherited in the search algorithms (**R6**). Such recommendation can be used as a general guide to any multi-objective test optimization problems at hand including uncertainty–wise test optimization.

The selection of a multiobjective search algorithm for dealing with a specific problem is also an important perspective that may have a huge impact on the performance of a search–based approach. Most of the existing works chose NSGA-II for solving multiobjective test optimization problems (e.g., Ref. [104] for test case prioritization, Ref. [108] for test case selection) and NSGA-II has proven to achieve promising results for addressing various multiobjective test optimization problems. However, several existing works have shown that Random-Weighted Genetic Algorithm (RWGA) can achieve better performance than NSGA-II, e.g., test set minimization problem [81]. Therefore, how to select the most appropriate multiobjective search algorithms largely depends on particular optimization problems and we would recommend applying and comparing at least NSGA-II (Pareto-based) and RWGA (weight-based) when solving a multiobjective test optimization problem when all the objectives are equally important (**R7**). However, as discussed before, if the objectives hold certain priories (i.e., user preferences), UPMOA [106] or RWGA is recommended to apply and compare since the traditional Pareto-based search algorithms (e.g., NSGA-II) cannot handle the multiobjective test optimization problems with user preferences (**R8**).

Furthermore, to assess the performance and justify the applicability of a multiobjective search algorithm, it is usually recommended to compare the chosen algorithms with at least one baseline algorithm for a sanity check (e.g., random search or greedy algorithm) (**R9**) [73,81,105,106,109,110]. Such practice can prove that the problem to be addressed at hand is complex enough, which requires the applications of an advanced multiobjective algorithm (e.g., NSGA-II). This recommendation is once again a generic one and applicable to any multiobjective test optimization problems and even single–objective test optimization problems.

Moreover, another important decision to make while employing multiobjective search algorithms lies in setting parameters for the selected algorithms. A typical practice is to use the default parameter settings that are, for example, provided by the jMetal library (e.g., crossover rate is 0.9, mutation probability is set as $1/n$, where n is the total number of variables (e.g., test cases) and population size is 100) [103,104,110]. Such default settings have often provided good results, and thus it is recommended to use at least the default settings for the selected multiobjective search algorithms (**R10**). Notice that parameter tuning is an alternate technique to find the best parameter settings for the multiobjective search algorithms, but parameter tuning is quite expensive in practice [74], which is considered as an optional recommendation if there are sufficient resources (e.g., manual effort) in practice for such parameter tuning activity. Interested readers can consult [74] for more details in terms of some parameter tuning techniques and key observations for search-based software engineering.

3.8.3 Evaluation Metrics

It is critical to employ proper evaluation metrics for evaluating the performance of multiobjective search algorithms along with the defined fitness function. The evaluation metrics can be chosen when different types of multiobjective search algorithms are applied, i.e., weight-based search algorithms and Pareto-based search algorithms. In terms of weight-based multiobjective search algorithms, it is recommended to directly use the obtained fitness function values as evaluation metrics for assessing the performance of algorithms (**R11**). Recall that weight-based search algorithms convert a multiobjective test optimization problem into a single-objective test optimization problem by assigning weights to each objective. When the search process is terminated, one fitness function value is produced that represents the quality of the corresponding solution. In addition, if users are also interested in particular individual objective (e.g., fault detection capability), it is also possible to compare the solutions obtained by search algorithms using each individual objective function value. However, notice that only the fitness function value can represent the overall quality of a solution obtained by a search algorithm.

With respect to the Pareto-based multiobjective search algorithms, it is recommended to employ standard quality indicators (e.g., hypervolume (HV), Epsilon (ϵ), Generalized Spread (GS), Generational Distance (GD)) as evaluation metrics to evaluate the performance of algorithms (**R12**). Our previous work [105] has proposed a practical guide in terms of selecting quality indicators for assessing the performance of Pareto-based multiobjective search algorithms based on theoretical foundations, literature review, and extensive empirical experiment, and thus it is always recommended to use this guide for choosing quality indicators when there is a need for Pareto-based search algorithm evaluation (**R13**). We briefly summarize the guide as below to make the book chapter self-contained. More details related to such guide can be consulted in Ref. [105]:

1. When a user is only concerned whether the obtained solutions are optimal or not (*Convergence*): If an ideal set of objective values (i.e., optimal objective values such as the best value for the objective of test minimization percentage is 0 when all the test cases are eliminated) is unknown for a multiobjective optimization problem, any of the quality indicators from GD, Euclidean Distance from the Ideal Solution (ED), ϵ, and Coverage (C) is recommended. On the opposite, ED is suggested if an ideal objective set is known before.

2. When a user is also concerned with the *Diversity* of the obtained solutions in addition to the convergence: When the number of objectives for a particular multiobjective optimization problem is more than six, Inverted Generational Distance (IGD) is recommended. Otherwise, HV is suggested if the number of objectives is less than or equal to six.

3. When a user prefers evaluating the performance of Pareto-based search algorithms from the two perspectives of *Convergence* and *Diversity* separately: Such situation may sometimes occur since separate quality indicators to measure convergence and diversity may be more accurate than the combined quality indicators, e.g., HV and IGD (mentioned as earlier). In this case, if the ideal objective set is known, ED together with Pareto front size (PFS) and GS is recommended. Otherwise, any of the quality indicators from GD, ED, ϵ, and C together with PFS and GS can be applied. It is worth mentioning that assessing the performance of Pareto-based multiobjective search algorithms from the two separate perspectives makes it difficult to draw a definite and clear conclusion which multiobjective search algorithm is better and thus we usually recommend selecting the combined quality indicators from HV or IGD with the aim of assessing the algorithm performance.

In addition, the running time taken by the search algorithms (both weight-based and Pareto-based multiobjective search algorithms) is also recommended to consider when evaluating the performance of algorithms since it would be practically infeasible to apply a search-based approach if it took the unacceptable running time to obtain optimal solutions (**R14**).

3.8.4 Statistical Tests

Recall that search algorithms are usually required to be run at least 10 times due to the random variations inherited and a large amount of data results will be obtained when the search process is finished. Thus, it is a common practice to employ standard statistical tests to analyze the results obtained and extracts the key observations. Notice that the following described statistical tests are generic and applicable to all the multiobjective search algorithms and any multiobjective optimization problems.

First, it is recommended to perform the descriptive statics (e.g., mean, median, and standard derivation) with the aim of acquiring direct impressions how good the results are (with the values of mean and median) and how the results are distributed (with the values of standard derivation) (**R15**). Moreover, we recommend performing the Vargha and Delaney statistics (A_{12}), which is a nonparametric effect size measure (**R16**). More specifically, in the context of SBSE, A_{12} is usually employed to measure the probability of yielding higher values for each objective function and fitness function value when comparing two multiobjective algorithms A and B. The two algorithms are considered to have equivalent performance if A_{12} is 0.5. The algorithm A has higher chances to obtain better solutions than the algorithm B if the value of A_{12} is greater than 0.5.

In addition, to compare the difference significance of several multiobjective search algorithms, it is recommended to perform a set of significance tests discussed as follows (**R17**). First, we recommend performing the Kruskal–Wallis test together with the Bonferroni correction [111] with the aim of determining if there are significant differences among multiple sets of results (samples) obtained by multiobjective search algorithms. Notice that the Bonferroni correction is used to adjust the P-value obtained by the Kruskal–Wallis test [111,112]. If the adjusted P-value of the Kruskal–Wallis test is more than 0.05, it indicates that there are no significant differences observed among the results obtained by various search algorithms and thus it can be concluded that the performance of search algorithms is statistically similar.

If the adjusted P-value is less than 0.05 that indicates that there exist significant differences among the results among by different search algorithms, the following statistical tests are recommended. First, the Shapiro–Wilk test [112] is recommended to determine the normality of obtained data samples with the aim of selecting an appropriate statistical test for checking the significance of the difference. The significance level is usually set as 0.05; i.e., a data sample is normally distributed if the P-value from the Shapiro–Wilk test is greater than 0.05. On the condition that the sample data is normally distributed, it is recommended to perform t-test (parametric test) [111] for determining whether there exist significant differences between the results obtained by different search algorithms. When the sample data is far away from a normal distribution, the Mann–Whitney U test (nonparametric test) [111] is suggested to determine the significance of the difference between the results obtained by different search algorithms. Similarly, the significance level is usually set as 0.05, i.e., there is a significant difference observed between the obtained results if the P-value is less than 0.05.

In addition, to assess the scalability of the multiobjective search algorithms, we recommend the Spearman's rank correlation coefficient (ρ) [111] to measure the correlations between the results obtained by the algorithms and the complexity of problems (**R18**). The value of ρ ranges from -1 to 1, i.e., a value greater than 0 denotes a positive correlation between the results of search algorithms and the complexity of problems, while a negative correlation is observed when the value of ρ is less than 0. A value of ρ close to 0 indicates that there is no correlation between the results obtained by search algorithms and the complexity of problems. Moreover, it is also recommended to report the significance of correlation using $Prob > |\rho|$, i.e., a value that is lower than 0.05 means that the observed correlation is statistically significant. For instance, with respect to addressing the test set minimization problem, our previous work [81] assessed the scalability of ten multiobjective search algorithms by measuring the correlation (using the Spearman's rank correlation coefficient) between the obtained results and the increasing complexity of problems measured by the number of features (i.e., testing functionalities of systems) included. The results showed that the performance of most of the multiobjective search algorithms was not significantly decreased when the problems become more complex when dealing with our test set minimization problem.

In summary, this section presents in total 18 recommendations (**Rs**) as shown in Table 3 that can be used by researchers and practitioners when there is a need to apply multiobjective search algorithms to tackle uncertainty–wise multiobjective test optimization problems.

4. THE STATE OF ART

This section presents the state of the art related to uncertainty–wise testing for CPSs, which includes a summary of the primary studies collected in a systematic mapping study that we conducted to review testing of CPSs under uncertainty/nondeterminism (Section 4.1) and a summary of the existing literature in terms of multiobjective test optimization (Section 4.2).

Table 3 Summary of the Recommendations

R #	Category	Description
1	Fitness function	It is recommended to first finalize cost–effectiveness measures (objective functions) before defining fitness function
2		It is recommended to define cost–effectiveness measures based on particular domains to fulfill practical requirements
3		It is recommended to directly use objective functions as fitness function when Pareto-based multiobjective algorithms are applied. If weight-based multiobjective algorithms are applied, it is recommended to assign weights to each objective function and convert the multiobjective test optimization problem to single–objective test optimization problem
4		It is recommended to limit the number of objectives less or equal to six if Pareto-based search algorithms are to be applied
5		It is recommended to apply Uniformly Distributed Weights (UDW) assignment strategy when there is a need to use weight-based genetic algorithms
6	Multiobjective search algorithms	It is recommended to repeat each multiobjective search algorithm at least 10 times for reducing the random variations inherited in search algorithms
7		It is recommended to apply and compare at least NSGA-II (Pareto-based) and RWGA (weight-based) when solving a multiobjective test optimization problem when all the objectives are equally important
8		It is recommended to apply and compare UPMOA or RWGA if the objectives hold certain priorities (i.e., user preferences)
9		It is recommended to compare the selected multiobjective search algorithms with at least one baseline algorithm for a sanity check (e.g., random search or greedy algorithm)
10		It is recommended to use at least the default settings for the selected multiobjective search algorithms

Table 3 Summary of the Recommendations—cont'd

R #	Category	Description
11	Evaluation metrics	It is recommended to directly use the obtained fitness function values as evaluation metrics for assessing the performance of weight-based multiobjective search algorithms
12		It is recommended to employ standard quality indicators (e.g., hypervolume (HV), Epsilon (ϵ)) as evaluation metrics to evaluate the performance of Pareto-based multiobjective search algorithms
13		It is recommended to use the guide from Ref. [105] for choosing quality indicators when there is a need for evaluating the performance of Pareto-based search algorithms.
14		It is recommended to compare the running time taken by the search algorithms when evaluating the performance of algorithms
15	Statistical tests	It is recommended to perform the descriptive statics (e.g., mean, median, and standard derivation) for acquiring direct impressions of the results
16		It is recommended to perform the Vargha and Delaney statistics (A_{12}), which is a nonparametric effect size measure for comparing which algorithm can have a higher chance of obtaining better results
17		When comparing the difference significance of several multiobjective search algorithms, it is recommended to first perform the Kruskal–Wallis test together with the Bonferroni correction. If the adjusted P-value is less than 0.05, the Shapiro–Wilk test is recommended to determine the normality of obtained data samples. If the data samples are normally distributed, t-test (parametric test) is recommended. Otherwise, the Mann–Whitney U test (nonparametric test) is recommended
18		It is recommended to perform the Spearman's rank correlation coefficient (ρ) to measure the correlations between the obtained results and the complexity of problems when assessing the scalability of multiobjective search algorithms

4.1 Summary of the Systematic Mapping

We conducted a systematic mapping study on testing CPSs under uncertainty by collecting and analyzing in total 26 research papers from the existing literature. In this section, we present the state of the art on uncertainty-wise testing for CPSs.

The work reported in Ref. [113] deals with time simulation methods that are applicable for testing protocol software embedded in communicating systems. Two types of nondeterminism are discussed in Ref. [113], which include: (1) internal nondeterminism that refers to "the non-deterministic implementation of the software under test" [114–116]; and (2) external nondeterminism that refers to "is triggered by the concurrent (and reactive) environment of the software" [117]. Internal nondeterminism can be handled through changing the implementation of the SUT to remove nondeterminism and test a set of serialization variants of the original implementation. To tackle external nondeterminism, simulation time is used to control the level of nondeterminism caused by the environment of the SUT. The test system is controlled in such a manner that external nondeterminism is allowed, reduced, or eliminated based on the instructions given by testers. To be more specific, external nondeterminism can be eliminated through simulation scheduling that serializes the communication between the test environment and the SUT. Simulation with deterministic discrete timing is used to enhance the controllability of testing while allowing several possible orders for the events. By allowing external nondeterminism, simulation time is used to scale timing in such a way that the tests can be executed in a host. The described methods have been evaluated in and applied to testing protocols for several standards.

In order to support black-box MBT of real-time embedded systems, the authors in Ref. [118] proposed a UML-based methodology to model the structure, behaviors, and constraints of the environment, where nondeterminism was intensively captured. To be more specific, intervals were applied to model timeout transitions. The unexpected behaviors of users could be captured through nondeterministic attributes whose legal values were constrained in OCL. For the nondeterminism of transitions, a probability distribution could be used. A set of stereotypes was proposed to capture the above-mentioned nondeterminism. The approach was applied to and assessed using two industrial case studies. Based on the environment models, the black-box model-based testing could be fully automated using search-based test case generation techniques and the generation of code simulating the environment.

The work in Ref. [119] extended the environment modeling methodology presented in Ref. [118] with a focus on the simulation for automated testing. A Java-based simulator was generated from the environment models using the model to text transformations. The nondeterminism was handled when generating the simulator by a nondeterministic engine that obtained the values of nondeterministic occurrences from the simulation configuration generated by the test framework. Therefore, the simulation becomes deterministic for each specific simulation configuration. The test framework aimed to find a simulation configuration that could lead to a system failure/error using search-based techniques [120,121]. Once such simulation configuration was found, a JUnit test case would be automatically generated, which is used to define two important components of the simulation including: (1) the configuration of the environment, e.g., number of sensors/actuators and their initialization; (2) the nondeterministic events in the simulation, e.g., variance in time-related events such as physical movements of hardware components, occurrence and type of hardware failures, and actions of the user(s). The proposed techniques were evaluated with three artificial problems and two industrial real-time embedded systems.

Using input from the environment models [118], Iqbal *et al.* focused on the selection of the test cases to facilitate the black-box testing of real-time embedded systems and investigated three test automation strategies, i.e., random testing (as a baseline), adaptive random testing, and genetic SBT in Ref. [122]. By random testing, test cases were randomly selected based on uniform probability. The basic idea of adaptive random testing was to reward diversity among test cases, as failing test cases usually tend to be clustered in contiguous regions of the input domain. A fitness function was defined to guide the search by estimating how close a test case can trigger a failure. The three approaches were validated using an industrial real-time embedded system, and the results showed that none of the three test automation strategies could fully dominate the others in all testing conditions. Moreover, the authors also provided practical guidelines in terms of applying these three techniques based on the experiment results.

In Ref. [123], a tool named UPPAAL-TRON [124] was proposed for online black-box conformance testing of real-time embedded systems based on nondeterministic timed automata specifications where nondeterministic action and timing were allowed. The notion of relativized timed input/output conformance was defined and it was possible to test the conformance of the system online based on the timed automata model and its environment assumptions that were explicitly captured. UPPAAL-TRON was implemented with a randomized online testing algorithm by extending the mature UPPAAL model checker. It could generate and execute tests event by event in real time through stimulating and monitoring the IUT (implementation under test). To perform these two functions, UPPAAL-TRON essentially computed the possible set of states symbolically based on the timed trace observed so far. The approach was further applied to test a rail-road intersection controller and was assessed in terms of error detection capability and computation performance. Moreover, some experience were shared in Ref. [124] with respect to applying UPPAAL-TRON in advanced electronic thermostat regulator that controls and monitors the temperature of industrial cooling plants.

Due to the lack of tool and techniques that could systematically tackle models capturing the indeterminacy as a result of concurrency, timing, and limited observability and controllability, David et al. [125] proposed a number of principles and algorithms of model-based test generation with UPPAAL tool support, aiming at efficiently testing the real-time embedded system under uncertainty. The methods proposed were complementary with each other in terms of the allowed uncertainty and observability.

Specifically, online testing is effective when the system is highly nondeterministic [123,124]. The main idea of online testing is to continually compute the possible set of states the model can occupy as test inputs/outputs or delays are observed. The uncertainty that the tester has about the possible state was thus captured by the state set. While testing a system with partial observability, testing based on partially observable games techniques, such as UPPAAL-POTIGA, a timed game solver for partially observable timed games [126] could be applied.

So as to devise deterministic test requirements yielding the maximum network stress test scenarios for testing real-time embedded systems, the precise timing information of the interactions, i.e., messages in a UML sequence diagram, was required [127]. However, such information was not always available, and thus timing uncertainty would impact the effectiveness of the generated test cases in terms of revealing real-time faults. In order to address timing uncertainty when generating test requirements, Garousi adapted the barrier scheduling heuristic and proposed a wait-notify stress test methodology (WNSTM) [127]. To increase the chances of discovering real-time faults originating from network traffic overloads, WNSTM generated stress test requirements by installing barriers via a counting semaphore before the maximum-stressing messages of each sequence diagram. WNSTM was further evaluated on a prototype distributed real-time system based on a real-world case study, and the results showed that WNSTM was effective in terms of detecting real-time faults.

Ali *et al.* tackled the test minimization problem in the context of uncertainty-wise testing [128]. The heuristics were defined to minimize cost (time) and maximize effectiveness (transition coverage) and uncertainty-related attributes that were measured using uncertainty theory. The goal of this work was to evaluate the impact of various mutation and crossover operators on the performance of NSGA-II in terms of solving the test minimization problem. Three mutation operators and three crossover operators, resulting in nine combinations for the setting of NSGA-II, were evaluated in a real CPS case study. The results showed that the Blend Alpha crossover operator together with the polynomial mutation operator showed the best performance.

Executable test cases could be generated from test ready models yielding to subjective uncertainties, as they were developed based on testers' assumptions about the expected behaviors of a CPS, its expected physical environment, and the potential future deployments. Zhang *et al.* [36] presented a model evolution framework (*UncerTolve*) to interactively improve the quality of test ready models, i.e., reduce the uncertainty, based on operational data. *UncerTolve* was characterized by three key features: (1) model execution was applied to validate the syntactic correctness and conformance of test ready models against real operational data; (2) the objective uncertainty measurements of test ready models, i.e., probabilities, were evolved via model execution; and (3) a machine learning technique was applied to evolve state invariants and guards of transitions. *UncerTolve* was evaluated using one industrial CPS case study and the results were promising.

A model-based testing framework for probabilistic systems was presented in Ref. [129], where both an offline algorithm and an online algorithm for test generation were proposed. The probabilistic input/output transition systems (pIOTSs) were employed to capture the probabilistic requirements of the SUT, and statistical methods were applied to assess if the frequencies of the observation during test execution conformed to the probabilities in the requirements according to the conformance relation defined in this paper. This framework was proved to be mathematically correct via the classical soundness and completeness properties.

As an extension to their previous work in Ref. [129], the authors presented key concepts of an MBT framework for probabilistic systems with continuous time in Ref. [130]. A solid core of a probabilistic test theory was provided to handle real-time stochastic behaviors, which were captured via Markov automata. Compared with their previous work, the novelty of this work was the inclusion of stochastic time and exponential delays.

A smart device is nondeterministic in nature due to many reasons, such as the inaccuracy in an analog measurement. In order to assess the reliability confidence in the smart device under nondeterminism, a black–box test environment was developed to automate the generation and execution of test data and the interpretation of the results in Ref. [131]. A finite state machine could be used to capture the system behavior, which was defined based on the behavior specified in the user manual. The nondeterminism of the smart device response was addressed by a result checker.

In order to assert the correctness of CPS with respect to extra–functional properties, a conformance testing approach, Timed Testing of a physical Quantity (TTQ), was proposed in Ref. [132]. TTQ checked the conformance of a running CPS via the timed model checker Uppaal and the timed online testing tool TRON, with respect to a formal timed automata description utilizing measurements of physical quantities subject to uncertainties. The physical measurements obtained from a running CPS would be aggregated and translated into a linear observer trace model (TM). TM and the formal description of the system behaviors, i.e., timed automata, were then jointly executed to check the reachability of the terminal location of the TM for inferring the conformance of the measurements with the expected behaviors of the system. In TTQ, the uncertainty was tackled in the sense that different intervals were employed for individual system modes of the hardware components. The proposed approach was evaluated for testing a wireless sensor node implementation with power measurements, and the effectiveness of the approach was highlighted by the experiments.

Conformance testing of CPS yielded to the state space explosion problem, as it depended on a reachability check that required a state space traversal based on the TM and timed automata. Therefore, Woehrle *et al.* proposed a segmented state space traversal approach for conformance testing of CPS in their later research [133]. This approach could improve the scalability of quantitative conformance testing of a CPS and it was demonstrated based on a case study of two communicating sensor nodes.

A test environment was presented in Ref. [134] in order to achieve: (1) easy management of test cases and (2) automatic testing. A repository was devised to manage system functions to be tested and system configurations to be tested separately. The test environment could automatically generate test cases with a template engine from certain system functions in PROMELA (PROcess Meta LAnguage) and system configurations in XML for the device under test. A modified SPIN (Simple PROMELA

Interpreter) model checker was applied to check certain properties of the PROMELA model where nondeterministic sequences were allowed. A prototype of an air conditioning system was developed for evaluating the proposed approach.

In the context of testing CPS product lines, the authors in Ref. [135] proposed a methodology with tool support (ASTERYSCO) to automatically generate simulation-based test system instances for testing individual configurations of CPSs. A test system to be generated was described in Simulink encapsulating several sources, e.g., test cases or test oracles. Feature model was applied to manage the variability of a CPS as well as the test system. The uncertainty was dealt by simulating the physical world with the context environment. Moreover, reactive test cases were supported so that they could monitor and react to the various states of the CPS.

CPS is stochastic in nature due to actuator inaccuracies, sensor readings, the rate of arrivals, component failure rates, etc., which could affect the functional correctness of a CPS. In order to detect system operating conditions yielding the worst system robustness, Abbas *et al.* introduced a robustness-guided testing for verifying stochastic CPS in Ref. [136], which could quantify how robustly a stochastic CPS satisfied a specification in Metric Temporal Logic. The goal was achieved by transforming the testing problem to an expected robustness minimization problem that was solved with stochastic optimization algorithms, i.e., Markov chain Monte Carlo algorithms.

In order to test embedded systems interacting with a continuous environment, i.e., hybrid systems, a model-based mutation testing approach was proposed in Ref. [137]. The discrete controller behaviors were modeled by classical action systems, whereas the evolutions of the environment were captured with qualitative differential equations. Nondeterminism was allowed in the action system model, such as internal actions and nondeterministic updates. The basic idea behind mutation testing was to mutate the system models and generate test cases capable of killing a set of mutated models. The generated test cases were then applied to execute the SUT and check the conformance between the mutations and SUT.

In order to test the safety properties in a CPS, the authors in Ref. [138] provided a formal framework for conformance testing based on a formal conformance relation to guarantee transference of safety properties. The conformance relation was named as each set conformance, which was a weaker relation than the existing trace conformance. Hybrid automata were applied to formally model the system behaviors. The conformance testing was based on computations and overapproximations of the reachable set, which also took care of bounded errors of simulations or real measurements. Moreover, a model-based input selection algorithm based on a reach set coverage measure was presented in order to reduce the number of tests for a given set of test cases. The framework was evaluated in the domain of autonomous driving, and the results showed that the conformance testing method could falsify more relations than the existing approaches.

A formal framework was provided in Ref. [139] for conformance testing of hybrid systems, a widely accepted mathematical model for many applications in embedded systems and CPSs. The goal of the conformance testing was to assert the conformance relation between the SUT and a formal specification using hybrid automata where nondeterminism was allowed in both continuous and discrete dynamics. Moreover, based on the notion of star discrepancy, a coverage measure was proposed for hybrid systems in order to: (1) quantify the validation completeness and (2) guide input stimulus generation. A test generation method (named as gRRT) with a prototype tool support was then proposed based on the rapidly exploring random tree algorithm for robotic planning guided by the coverage measure. gRRT was applied to a number of analog circuits and control applications. This test generation algorithm was later refined in Ref. [140] guided by both the coverage measure and the property to verify. In addition, the partial observability problem in test execution was also addressed in Ref. [140] through an estimation of the current location and the continuous state of the SUT with a hybrid observer.

Self-adaptation aims to address the challenges in CPSs capable of changing its behavior as well as structure in response to changes in the operating environment/context. Testing such self-adaptive behaviors is challenging due to the infinite reaction loop and uncertain interaction. A novel approach named as SIT (Sample-based Interactive Testing) was presented in Ref. [141] to test self-adaptive apps in an efficient and light-weight way. The input space of a self-adaptive app could be systematically split, adaptively explored, and mapped to the testing of different behaviors by an interactive app model and a test generation technique. The impact of environmental constraints and uncertainty on an app's input/output pairs was captured in the interactive app model, enabling a systematic and guided exploration of its space. The test generation techniques only sampled inputs required by the exploration. Moreover, external uncertainty due to the errors in measuring environmental conditions was also handled in the SIT approach. An uncertainty specification was defined based on a set of functions mapping a given environment's output parameter to its corresponding app's input parameters together with the associated error range. The effectiveness and efficiency of SIT were evaluated with real-world self-adaptive apps.

Ramirez *et al.* proposed a search-based approach named as Loki to automatically explore environmental conditions capable of revealing requirements violations and latent behaviors in a dynamic adaptive system [142]. To be more specific, Loki first generated configurations specifying the type, duration, and severity of noise for each sensor in a dynamic adaptive system. Second, Loki simulated the system and environmental conditions according to the generated configurations. During simulation, a set of utility functions were applied to assess the satisfaction of the system requirement captured with goal-based models and guide the search toward those environmental conditions producing the most distinct behaviors. Loki was applied to an autonomous vehicle system to illustrate several examples of requirement violations and latent behaviors.

Operating contexts for dynamic adaptive systems constitute main uncertainty that may affect the system behaviors at a run time. An evolutionary computation-based approach (FENRIR) was proposed in Ref. [143]. FENRIR could automatically explore how the varying operational contexts affected a dynamic adaptive system with instrumented code by searching for the system and environmental parameters that could produce previously unexamined system execution traces. The resulting execution traces could facilitate the identification of potential bugs in the code. FENRIR was evaluated on an industry-provided problem, management of a remote data mirroring network. Experiment results showed that FENRIR could provide a significantly greater coverage of execution paths as compared with randomized testing. Moreover, in their following research, Fredericks *et al.* [144] further proposed an evolutionary computation-based approach (named Veritas) to adapt requirement-based test cases at run time in response to different system and environmental conditions. Specifically, Veritas monitored the environment to collect evidence of changes in the environment and adapted individual test cases to ensure testing relevance that was assessed with utility functions. An online evolutionary algorithm (i.e., $(1+1)$-ONLINE EA) was then applied to facilitate run-time adaptation of test case parameters once required. Veritas was evaluated using an intelligent robotic vacuum, and the results showed that Veritas could largely reduce the mismatch between test cases and operational context caused by uncertainty within the system and the environment.

4.2 Multiobjective Test Optimization

Many software testing problems are multiobjective in nature (e.g., test set minimization problem [81,109]) and the existing literature mainly studied the multiobjective test optimization from three perspectives, i.e., test case selection, test set minimization, and test case prioritization [80,106,108,145–147]. *Test Case Selection* aims at selecting a set of relevant test cases from the entire test set for cost-effectively testing the systems or programs under test. *Test Set Minimization* minimizes a given test set for eliminating redundant test cases with the aim to reduce the total cost of testing (e.g., execution time) while preserving high effectiveness (e.g., fault detection capability). *Test Case Prioritization* prioritizes a test set into an optimal order for detecting faults as early as possible while satisfying other predefined criteria, e.g., reducing the execution time of test cases. It is worth mentioning that it is practically infeasible to only take a single objective into account when studying the above-mentioned three testing problems [145]. Thus, researchers usually focus on proposing, defining, and formulating a set of cost-effective objectives based on specific contexts and domains (e.g., telecommunication [81,109]), which are further taken as input for guiding the search toward obtaining trade-off solutions that can balance different criteria [71,80,103,104, 106,108,145,148,149].

The principle foundation of SBST is to formulate and encode various multiobjective software testing problems into multiobjective test optimization problems that can be efficiently solved using various multiobjective search algorithms (e.g., nondominated sorting genetic algorithm II (NSGA-II) [72]). According to a publicly available SBSE repository maintained by the CREST center [150], in total 821 research papers have been focusing on tackling a variety of testing challenges, and out of these 821 works, a large number of multiobjective test problems have been increasingly researched, e.g., test case selection problem [103,108] and test case prioritization [71,104,106]. Harman *et al.* have conducted a systematic literature review for SBSE [148] that described and sketched a variety of SE applications employing search algorithms to address multiobjective test optimization problems. In particular, a set of objectives have been listed in Ref. [145] (e.g., fault detection capability, code coverage), which can be used for multiobjective test optimization in the context of software testing and have been researched in the state of the art [80].

For instance, in terms of test case selection, two cost-effective objectives (i.e., code coverage, execution time) have been defined with the aim of selecting optimal test cases in Ref. [151]. The two-objective problem has been further converted into a single-objective problem by assigning weights to each objective followed by solving the problem using search algorithms [151]. Yoo and Harman [108] formally defined three cost–effectiveness measures (i.e., code coverage, fault detection history, and execution time) in the context of regression testing and employed a greedy algorithm and a multiobjective search algorithm NSGA-II for selecting optimal test cases from a given test set. Our previous work [103] proposed to consider the importance of test cases, the potential impact caused by test case failure, and the likelihood of detecting faults by test cases into account when performing test case selection. We further empirically evaluated eight existing multiobjective search algorithms (e.g., NSGA-II) with two weight assignment strategies (i.e., fixed-weight strategy and randomly assigned weight strategy) using one industrial case study and a large number of artificial problems. With respect to test set minimization, our previous work [81] attempted to deal with minimization problem by defining five cost–effectiveness measures (e.g., test minimization percentage, feature pairwise coverage, and fault detection capability). We further empirically evaluated the performance of eight multiobjective search algorithms (including NSGA-II and RWGA) in conjunction with our defined fitness function using 1 industrial case study and 500 artificial problems. An improved search algorithm was proposed in Ref. [106] with the aim of incorporating predefined user preferences (based on different domains) when solving multiobjective test optimization problems (e.g., test set minimization problem). As for test case prioritization, a time–aware test case prioritization problem has been addressed in Ref. [152] by employing linear integer programming with the aim of prioritizing test cases into an optimal order with a given time budget. Our previous work [104] studied the test case prioritization problem by proposing four effectiveness measures (e.g., configuration coverage, test API coverage, fault detection capability) in the context of telecommunication domain. Two prioritization strategies were further defined and incorporated into the fitness function, which include: (1) *Incremental Unique Coverage* that only considers the incremental unique elements (e.g., test APIs) covered by the test case as compared with the ones covered by the prioritized test cases; and (2) *Position Impact* that gives more impact to a test case with a higher execution position (i.e., scheduled to be executed earlier).

However, to the best of our knowledge, there are only a few works [33,77] in the literature (as discussed in Section 4.1) that addressed the uncertainty-wise multiobjective test optimization problem that explicitly takes uncertainty into account. Thus, there is still a large open research room that requires further investigation in terms of this angle.

5. CONCLUSION AND FUTURE RESEARCH DIRECTIONS

Due to the fact that uncertainty is unavoidable in the behaviors of CPSs, it is important that uncertainty in CPSs both in its internal behavior and in its physical environment must be considered explicitly during their verification and validation. In this direction, this chapter introduced uncertainty-wise testing in the context of CPSs. More specifically, we summarized our existing works from the angles of Uncertainty-wise Model-Based Testing, Uncertainty-wise Test Ready Model Evolution, and Uncertainty-wise Multiobjective Test Optimization, in addition to providing potential future research directions. Moreover, we also provided a survey of the existing uncertainty-wise testing approaches to present where the current state-of-the-art stands in this area.

Given the anticipated substantial rise in CPS applications in the future in both daily lives and critical domains, it will become even more critical for CPSs to handle uncertainty in a reliable way. This opens up a lot of new research directions for uncertainty-wise testing in the future. We summarize some research directions in the area of uncertainty-wise testing below:

First, existing uncertainty-wise testing solutions support only testing functional behaviors in the presence of environment uncertainty with a limited extent as we discussed in the related work section. In this direction, new research is required to define novel testing methods for testing extra-functional behaviors of CPSs, e.g., related to security and safety, under various types of uncertainties. Such testing methods will ensure that extra-functional requirements (e.g., safety and security requirements) are not violated under various types of uncertainties. Note that if extra-functional requirements are not violated in nominal conditions, it does not mean that such requirements will not be violated under uncertainty and thus it is important that future research in uncertainty-wise testing shall also focus on testing extra-functional requirements in the presence of uncertainty.

Second, due to the fact that a large number of CPSs interact with humans, testing methods must take into consideration uncertainty introduced by humans into CPS operations. Given that understanding human behavior requires expertise from the domain of human psychology, such testing methods shall incorporate knowledge from the domain of human psychology to be considered explicitly in all the phases of testing. In addition to uncertainty-related theories, such as Uncertainty and Probability theories, testing solutions shall also be grounded on theories related to study of human behaviors in uncertainty such as Prospect theory [153]. Needless to say, such testing solutions shall be interdisciplinary requiring expertise from the domains of ICT and Human Psychology with the eventual implementations as novel testing tool suites.

Third, the future modeling solutions to support automated testing shall provide integrated solutions incorporating modeling expected behavior of software, hardware, and physical environment including humans. In addition, the modeling solutions must provide an implementation of relevant uncertainty-related theories, e.g., Uncertainty Theory and Probability Theory, and theories from other domains such as Prospect theory from human psychology.

Fourth, the current modeling tools for system design and also for model-based testing only support checking the syntactic correctness of models; however, none of the existing modeling tools support automated validation of test models. This opens up a new research direction for devising new methods to validate test ready models of CPSs capturing uncertainty. In terms of testing, it is important that test ready models are semantically correct; otherwise, test cases generated from such test ready models will not be correct. In the presence of uncertainty, such validation methods become even more complicated because of lack of knowledge or missing information as compared to the situation where test ready models do not capture uncertainty.

Fifth, novel test strategies for uncertainty-wise testing of CPSs are required to be developed in the future. Such test strategies shall not only focus on finding optimized test paths in test ready models but also focus on finding best test data including uncertainty test data, to find faults cost-effectively. Notice that with uncertainty-wise test strategies, we are looking for faults that could only be observed in the presence of uncertainty, thus requiring the development of a potentially new classification of uncertainty-wise faults for CPSs. Such a classification can further facilitate the development of new mutation operators that can be used to assess the cost–effectiveness of newly defined uncertainty-wise test generation strategies and support building the foundations of mutation testing for uncertainty-wise testing. Finally, in terms of test optimization, we foresee the need for the development of new search algorithms for both single-objective and multiobjective optimization such as extensions to Genetic Algorithms (GAs) and Nondominated Sorting Genetic Algorithm II (NSGA-II). Such algorithms shall be extended to incorporate uncertainty in their inputs to provide optimized solutions even when faced with uncertain inputs.

Sixth, as with any area of research, more empirical evaluations are required for uncertainty-wise testing. This includes not only the empirical evaluations in academic settings but also more applications in diverse domains to determine which uncertainty-wise techniques are better under which contexts.

Finally, as we discussed that there is no existing standard for modeling uncertainty and recently a new Uncertainty Modeling [29] standard is initiated at the OMG. This standard will be a generic uncertainty-modeling standard and will probably require tailoring for specialized domains and applications. In addition, more standardization efforts are expected to integrate uncertainty-related aspects into existing software and systems modeling languages such as SysML [54] and MARTE [55], in addition to testing standards, such as the UML Testing Profile [52].

ACKNOWLEDGMENTS

This research was supported by the EU Horizon 2020 funded project (Testing Cyber-Physical Systems under Uncertainty, Project No. 645463). T.Y and S.A. are also supported by RCN funded Zen-Configurator project, RFF Hovedstaden funded MBE-CR project, RCN funded MBT4CPS project, RCN funded Certus SFI, and the EU COST action MPM4CPS. H.L. is funded by RCN funded Zen-Configurator project, S.W. is funded by RFF Hovedstaden funded MBE-CR project and RCN funded Certus SFI, and M.Z. is funded by the EU Horizon 2020 funded project (Testing Cyber-Physical Systems under Uncertainty, Project No. 645463).

 APPENDIX

List of Standardization Bodies and Standards With Uncertainty

Body	Modeling/ Testing/ MBT/Other	Standard	Uncertainty
OMG	Modeling	Unified Modeling Language (UML) [53]	No
		UML Profile for MARTE: Modeling and Analysis of Real-Time and Embedded Systems [55]	Yes (Probability)
		Systems Modeling Language (SysML) [54]	Yes (Probability)
		Object Constraint Language (OCL) [56]	No
		MetaObject Facility (MOF) [57]	No
		Structured Assurance Case Metamodel (SACM) [154]	Yes (Evidence, Confidence, Confidence Level)
	MBT	UML Testing Profile (UTP) [52]	No

List of Standardization Bodies and Standards With Uncertainty—cont'd

Body	Modeling/ Testing/ MBT/Other	Standard	Uncertainty
ISO, IEC and IEEE	Modeling	ISO/IEC 10746 series—Reference Model of Open Distributed Processing (RM-ODP) [61]	No
		ISO/IEC 19505 series—OMG UML 2.4.1 [58]	No
		ISO/IEC 19507:2012—OMG OCL 2.3.1 [59]	No
		ISO/IEC 19506:2012—OMG Architecture-Driven Modernization (ADM)— Knowledge Discovery Meta-Model (KDM) [60]	No
	Testing	ISO/IEC/IEEE 29119 series— Software Testing Standard [64]	No
		ISO/IEC 9646 series—Open Systems Interconnection (OSI)— Conformance Testing Methodology and Framework [65]	No
		IEEE 1012-2012—System and Software Verification and Validation [66]	No
		IEEE 829-2008—Software and System Test Documentation [67]	No
		IEEE SA-1008-1987—IEEE Standard for Software Unit Testing [68]	No
		IEEE 1044-2009—Classification for Software Anomalies [69]	No
	Other	ISO/IEC/IEEE 24765:2010— Systems and Software Engineering—Vocabulary [155]	No
		ISO/IEC/IEEE 42010:2011— Systems and Software Engineering—Architecture Description [156]	No

Continued

List of Standardization Bodies and Standards With Uncertainty—cont'd

Body	Modeling/ Testing/ MBT/Other	Standard	Uncertainty
		ISO/IEC/IEEE 15288:2015— Systems and Software Engineering—System Life Cycle Processes [157]	No
		ISO/IEC 16085:2006—Systems and Software Engineering—Life Cycle Processes—Risk Management [158]	No
		ISO/IEC 25010:2011—Systems and Software Quality Requirements and Evaluation (SQuaRE)—System and Software Quality Models [159]	No
		ISO/IEC 15026 series—Systems and Software Assurance [160]	No
		ISO/IEC 12207:2008—Systems and Software Engineering— Software Life Cycle Processes [161]	No
		ISO/IEC Guide 98 series— Uncertainty of Measurement [162]	Yes
		ISO/IEC 10165-7:1996—Open Systems Interconnection (OSI)— Structure of Management Information: General Relationship Model [163]	No
		IEC Guide 115:2007—Application of Uncertainty of Measurement to Conformity Assessment Activities in the Electrotechnical Sector [164]	Yes
		IEC 61508:2010—Functional Safety of Electrical/Electronic/ Programmable Electronic Safety- Related Systems [70]	Yes (Probability)
		IEC 31010:2009—Risk Assessment Techniques [165]	No

List of Standardization Bodies and Standards With Uncertainty—cont'd

Body	Modeling/ Testing/ MBT/Other	Standard	Uncertainty
		IEEE 730-2014—Software Quality Assurance Processes [166]	No
		IEEE 1061-1998—Software Quality Metrics Methodology [167]	No
		IEEE P2413—Standard for an Architectural Framework for the Internet of Things (IoT) [168]	No
		ISO 9000 series—Quality Management [169]	No
		ISO 31000—Risk Management [170]	Yes (Risk, Uncertainty, Effect, Likelihood)
		ISO 3534-1:2006—General Statistical Terms and Terms Used in Probability [171]	Yes
		ISO 21748:2010—Guidance for the Use of Repeatability, Reproducibility, and Trueness Estimates in Measurement Uncertainty Estimation [172]	Yes
		ISO/TR 13587:2012—Three Statistical Approaches for the Assessment and Interpretation of Measurement Uncertainty [173]	Yes
		ISO/TS 17503:2015—Guidance on Evaluation of Uncertainty Using Two-Factor Crossed Designs [174]	Yes
		ISO 9241 series—Ergonomics of Human–System Interaction [175]	No
JCGM	Other	JCGM 200:2012—International Vocabulary of Metrology—Basic and General Concepts and Associated Terms (VIM) [176]	Yes

Continued

List of Standardization Bodies and Standards With Uncertainty—cont'd

Body	Modeling/ Testing/ MBT/Other	Standard	Uncertainty
ETSI	Testing	ETSI TR 102 422 V1.1.1 (2005-04)—IMS Network Integration Testing Infrastructure Testing Methodology [177]	No
		ETSI EG 203 130 V1.1.1 (2013-04)—Methodology for Standardized Test Specification Development [178]	No
		ETSI TR 101 583 V1.1.1 (2015-03)—Security Testing; Basic Terminology [179]	No
		ETSI ES 201 873 series on TTCN-3 [180]	No
		ETSI ES 203 119 series on Test Description Language (TDL) [181]	No
	Testing, MBT	ETSI TR 102 840 V1.2.1 (2011-02)—Methods for Testing and Specifications (MTS); Model-Based Testing in Standardization [182]	No
		ETSI ES 202 951 V1.1.1 (2011-07)—Methods for Testing and Specification (MTS); Model-Based Testing (MBT); Requirements for Modeling Notations [62]	No
		ETSI EG 201 015 V2.1.1 (2012-02)—Methods for Testing and Specification (MTS); Standards Engineering Process; a Handbook of Validation Methods [63]	No
OASIS	Other	Open Services for Life Cycle Collaboration (OSLC) [183]	No

REFERENCES

[1] D.B. Rawat, J.J. Rodrigues, I. Stojmenovic, Cyber-Physical Systems: From Theory to Practice, CRC Press, United States, 2015.

[2] S. Sunder, Foundations for innovation in cyber-physical systems, in Proceedings of the NIST CPS Workshop, Chicago, IL, USA, 2012.

[3] E. Geisberger, M. Broy, Living in a Networked World: Integrated Research Agenda Cyber-Physical Systems (agendaCPS), Herbert Utz Verlag, Munich, Germany, 2015.

[4] S. Hangal, M.S. Lam, in: Tracking down software bugs using automatic anomaly detection, Proceedings of the 24th International Conference on Software Engineering (ICSE 2002), 2002, pp. 291–301.

[5] S. Anand, E.K. Burke, T.Y. Chen, J. Clark, M.B. Cohen, W. Grieskamp, M. Harman, M.J. Harrold, P. McMinn, A. Bertolino, J. Jenny Li, H. Zhu, An orchestrated survey of methodologies for automated software test case generation, J. Syst. Softw. 86 (8) (2013) 1978–2001. http://dx.doi.org/10.1016/j.jss.2013.02.061.

[6] C. Nie, H. Leung, A survey of combinatorial testing, ACM Comput. Surv. 43 (2) (2011) 11.

[7] P. McMinn, Search-based software test data generation: a survey, Softw. Test. Verif. Reliab. 14 (2) (2004) 105–156.

[8] S. Ali, L.C. Briand, H. Hemmati, R.K. Panesar-Walawege, A systematic review of the application and empirical investigation of search-based test case generation, IEEE Trans. Softw. Eng. 36 (6) (2010) 742–762.

[9] J. Burnim, K. Sen, in: Heuristics for scalable dynamic test generation, 23rd IEEE/ACM International Conference on Automated Software Engineering (ASE), 2008, pp. 443–446.

[10] L. Cseppento, Z. Micskei, in: Evaluating symbolic execution-based test tools, 2015 IEEE 8th International Conference on Software Testing, Verification and Validation (ICST), 2015, pp. 1–10.

[11] J. Offutt, A. Abdurazik, Generating tests from UML specifications, in: International Conference on the Unified Modeling Language, 1999, pp. 416–429.

[12] J. Offutt, S. Liu, A. Abdurazik, P. Ammann, Generating test data from state-based specifications, Softw. Test. Verif. Reliab. 13 (1) (2003) 25–53.

[13] M. Utting, A. Pretschner, B. Legeard, A taxonomy of model-based testing approaches, Softw. Test. Verif. Reliab. 22 (5) (2012) 297–312.

[14] M. Harman, P. McMinn, A theoretical and empirical study of search-based testing: local, global, and hybrid search, IEEE Trans. Softw. Eng. 36 (2) (2010) 226–247. http://dx.doi.org/10.1109/TSE.2009.71.

[15] M. Utting, B. Legeard, Practical Model-Based Testing: A Tools Approach, Morgan Kaufmann, San Francisco, CA, USA, 2010.

[16] T. Yue, S. Ali, M. Zhang, in: RTCM: a natural language based, automated, and practical test case generation framework, Proceedings of the 2015 International Symposium on Software Testing and Analysis, 2015, pp. 397–408.

[17] S. Ali, M. Zohaib Iqbal, A. Arcuri, L.C. Briand, Generating test data from OCL constraints with search techniques, IEEE Trans. Softw. Eng. 39 (10) (2013) 1376–1402.

[18] M. Mitchell, An Introduction to Genetic Algorithms, The MIT Press, Cambridge, 1996.

[19] EvoSuite—Automatic Test Suite Generation for Java, Accessed 2017, http://www.evosuite.org/.

[20] M. Zhang, B. Selic, S. Ali, T. Yue, O. Okariz, R. Norgren, in: Understanding uncertainty in cyber-physical systems: a conceptual model, Proceedings of the 12th European Conference on Modelling Foundations and Applications (ECMFA), 2016, pp. 247–264.

[21] M. Zhang, S. Ali, T. Yue, R. Norgre, An Integrated Modeling Framework to Facilitate Model-Based Testing of Cyber-Physical Systems under Uncertainty, Technical Report 2016-02, Simula Research Laboratory, 2016. https://www.simula.no/publications/integrated-modeling-framework-facilitate-model-based-testing-cyber-physical-systems.

[22] M. Zhang, T. Yue, S. Ali, B. Selic, O. Okariz, R. Norgren, K. Intxausti, Specifying Uncertainty in Use Case Models in Industrial Settings, Technical Report 2016-15, Simula Research Laboratory, 2016. https://www.simula.no/publications/specifying-uncertainty-use-case-models-industrial-settings.

[23] U-RUCM: Specifying Uncertainty in Use Case Models, Accessed 2016, http://zentools.com/rucm/U_RUCM.html.

[24] H. Zhang, T. Yue, S. Ali, C. Liu, in: Facilitating requirements inspection with search-based selection of diverse use case scenarios, BICT'15 Proceedings of the 9th EAI International Conference on Bio-Inspired Information and Communications Technologies (Formerly BIONETICS), 2015, pp. 229–236.

[25] T. Yue, L.C. Briand, Y. Labiche, aToucan: an automated framework to derive UML analysis models from use case models, ACM Trans. Softw. Eng. Methodol. 24 (3) (2015) 13.

[26] T. Yue, L.C. Briand, Y. Labiche, Facilitating the transition from use case models to analysis models: approach and experiments, ACM Trans. Softw. Eng. Methodol. 22 (1) (2013) 1–38 (Article No. 5).

[27] T. Weilkiens, Systems Engineering With SysML/UML: Modeling, Analysis, Design, Morgan Kaufmann, United States, 2011.

[28] S. Friedenthal, A. Moore, R. Steiner, A Practical Guide to SysML: The Systems Modeling Language, Morgan Kaufmann, United States, 2014.

[29] Uncertainty Modeling (UM), Accessed 2016, http://www.omgwiki.org/uncertainty/doku.php?id=start.

[30] H. Song, D.B. Rawat, S. Jeschke, C. Brecher, Cyber-Physical Systems: Foundations, Principles and Applications, Morgan Kaufmann, United States, 2016.

[31] C. Talcott, Cyber-physical systems and events, in: M. Wirsing, J.-P. Banâtre, M. Hölzl, A. Rauschmayer (Eds.), Software-Intensive Systems and New Computing Paradigms: Challenges and Visions, Springer, Berlin, Heidelberg, 2008, pp. 101–115.

[32] S. Ali, L.C. Briand, M.J.-U. Rehman, H. Asghar, M.Z.Z. Iqbal, A. Nadeem, A state-based approach to integration testing based on UML models. Inf. Softw. Technol. 49 (11–12) (2007) 1087–1106. http://dx.doi.org/10.1016/j.infsof.2006.11.002.

[33] M. Zhang, S. Ali, T. Yue, M. Hedman, Uncertainty-Based Test Case Generation and Minimization for Cyber-Physical Systems: A Multi-Objective Search-Based Approach, Technical Report 2016-13, Simula Research Laboratory, 2016. https://www.simula.no/publications/uncertainty-based-test-case-generation-and-minimization-cyber-physical-systems-multi.

[34] L. Briand, Y. Labiche, in: A UML-based approach to system testing, International Conference on the Unified Modeling Language, 2001, pp. 194–208.

[35] A. Abdurazik, J. Offutt, in: Using UML collaboration diagrams for static checking and test generation, International Conference on the Unified Modeling Language, 2000, pp. 383–395.

[36] M. Zhang, S. Ali, T. Yue, R. Norgre, Uncertainty-wise evolution of test ready models, Inform. Softw. Technol. (2017). http://dx.doi.org/10.1016/j.infsof.2017.03.003.

[37] M. Zhang, S. Ali, T. Yue, R. Norgren, Interactively Evolving Test Ready Models With Uncertainty Developed for Testing Cyber-Physical Systems, Technical Report 2016-12, Simula Research Laboratory, 2016. https://www.simula.no/

publications/interactively-evolving-test-ready-models-uncertainty-developed-testing-cyber-physical.

[38] P. Samuel, R. Mall, A.K. Bothra, Automatic test case generation using unified modeling language (UML) state diagrams, IET Softw. 2 (2) (2008) 79–93.

[39] L.C. Briand, Y. Labiche, Y. Wang, Using Simulation to Empirically Investigate Test Coverage Criteria Based on Statechart, in: Proceedings of the 26th International Conference on Software Engineering, 2004, pp. 86–95.

[40] W. Feller, An Introduction to Probability Theory and Its Applications, John Wiley & Sons, United States, 2008.

[41] B. Liu, Uncertainty Theory, Springer, Germany, 2015.

[42] C.C. Michael, G. McGraw, M.A. Schatz, Generating software test data by evolution, IEEE Trans. Softw. Eng. 27 (12) (2001) 1085–1110.

[43] A.J. Offutt, J. Pan, Automatically detecting equivalent mutants and infeasible paths, Softw. Test. Verif. Reliab. 7 (3) (1997) 165–192.

[44] M. Harman, K. Lakhotia, P. McMinn, in: A multi-objective approach to search-based test data generation, Proceedings of the 9th Annual Conference on Genetic and Evolutionary Computation, 2007, pp. 1098–1105.

[45] GraphWalker, Accessed 2017, http://graphwalker.github.io/.

[46] MoMuT, Accessed 2017, https://www.momut.org/.

[47] OSMO MBT Tool, Accessed 2017, https://github.com/mukatee/osmo.

[48] TestCast MBT—Model Based Testing Platform. Accessed 2007, http://www.elvior.com/en.

[49] Conformiq Creator & Designer, Accessed 2017, https://www.conformiq.com/products/.

[50] MBTsuite—The Testing Framework, Accessed 2010, http://www.mbtsuite.com/home.html.

[51] National Standards Bodies, Accessed 2015, http://www.nist.gov/iaao/stnd-org.cfm.

[52] OMG, UML Testing Profile, http://www.omg.org/spec/UTP/, 2013.

[53] OMG, Unified Modeling Language (UML), http://www.omg.org/spec/UML/, 2015.

[54] OMG, System Modeling Language, http://www.omg.org/spec/SysML/, 2014.

[55] OMG, UML Profile For MARTE: Modeling and Analysis of Real-Time Embedded Systems, http://www.omg.org/spec/MARTE/, 2011.

[56] OMG, Object Constraint Language (OCL), http://www.omg.org/spec/OCL/, 2014.

[57] OMG, Meta Object Facility (MOF) Core Specification (Version 2.4.2), http://www.omg.org/spec/MOF/2.4.2, 2014.

[58] ISO/IEC, ISO/IEC 19505-x: Information Technology—Object Management Group Unified Modeling Language (OMG UML), 2012.

[59] ISO/IEC, ISO/IEC 19507: Information Technology—Object Management Group Object Constraint Language (OCL), 2012.

[60] ISO/IEC, ISO/IEC 19506: 2012 Information technology—Object Management Group Architecture-Driven Modernization (ADM)—Knowledge Discovery Meta-Model (KDM), http://www.iso.org/iso/catalogue_detail.htm?csnumber=32625, 2012.

[61] ISO/IEC, ISO/IEC 10746-x: Information Technology— Open Distributed Processing, 2009).

[62] ETSI, ETSI ES 202 951 Methods for Testing and Specification (MTS); Model-Based Testing (MBT); Requirements for Modelling Notations (v1.1.1), http://www.etsi.org/deliver/etsi_es/202900_202999/202951/01.01.01_60/es_202951v010101p.pdf, 2011.

[63] ETSI, ETSI EG 201 015 Methods for Testing and Specification (MTS); Standards Engineering Process; a Handbook of Validation Methods (v1.2.1), http://www.

etsi.org/deliver/etsi_eg/201000_201099/201015/02.01.01_60/eg_201015v020101p.
pdf, 2012.

[64] ISO/IEC/IEEE, ISO/IEC/IEEE 29119-x: Software Testing, http://www.
softwaretestingstandard.org/, 2013.

[65] ISO/IEC, ISO/IEC 9646-x: Information technology—Open Systems Interconnection—
Conformance testing methodology and framework, 1994).

[66] IEEE, IEEE standard for system and software verification and validation, in: IEEE Std
1012-2012 (Revision of IEEE Std 1012-2004), 2012, pp. 1–223. http://ieeexplore.
ieee.org/xpl/articleDetails.jsp?arnumber=6204026.

[67] IEEE, IEEE standard for software and system test documentation—redline, in: IEEE
Std 829-2008 (Revision of IEEE Std 829-1998)—Redline, 2008, pp. 1–161.

[68] IEEE, IEEE standard for software unit testing, in: ANSI/IEEE Std 1008-1987, 1986.
p. 0_1. http://ieeexplore.ieee.org/xpl/articleDetails.jsp?arnumber=27763.

[69] IEEE, IEEE standard classification for software anomalies, in: IEEE Std 1044–2009
(Revision of IEEE Std 1044-1993), 2010, pp. 1–23. http://ieeexplore.ieee.org/xpl/
articleDetails.jsp?arnumber=5399061.

[70] IEC, IEC 61508: Commented Version, Functional Safety of Electrical/Electronic/
Programmable   Electronic   Safety-Related   Systems,   http://www.iec.ch/
functionalsafety/standards/, 2010.

[71] S. Wang, S. Ali, T. Yue, Ø. Bakkeli, M. Liaaen, in: Enhancing test case prioritization
in an industrial setting with resource awareness and multi-objective search, Proceedings of the 38th International Conference on Software Engineering Companion (ICSE
'16), 2016, pp. 182–191.

[72] K. Deb, A. Pratap, S. Agarwal, T. Meyarivan, A fast and elitist multiobjective genetic
algorithm: NSGA-II, IEEE Trans. Evol. Comput. 6 (2) (2002) 182–197, http://dx.doi.
org/10.1109/4235.996017.

[73] A. Arcuri, L. Briand, in: A practical guide for using statistical tests to assess randomized
algorithms in software engineering, Proceedings of the 33rd International Conference
on Software Engineering, Waikiki, Honolulu, HI, USA, 2011, pp. 1–10.

[74] A. Arcuri, G. Fraser, in: On parameter tuning in search based software engineering,
International Symposium on Search Based Software Engineering (SSBSE), 2011.

[75] S. Ali, T. Yue, Evaluating normalization functions with search algorithms for solving
OCL constraints, Testing Software and Systems: 26th IFIP WG 6.1 International
Conference, ICTSS 2014, Madrid, Spain, September 23–25, 2014. in: M.G. Merayo,
E.M. de Oca (Eds.), Proceedings, Springer, Berlin, Heidelberg, 2014, pp. 17–31.

[76] S. Ali, T. Yue, M. Zhang, Tackling uncertainty in cyber-physical systems with automated testing, ADA User J. 37 (4) (2016) 219–222.

[77] S. Ali, Y. Li, T. Yue, M. Zhang, Uncertainty-Wise and Time-Aware Test Case
Prioritization With Multi-Objective Search, Technical Report 2017-03, Simula
Research Laboratory, 2017. https://www.simula.no/publications/uncertainty-wise-
and-time-aware-test-case-prioritization-multi-objective-search.

[78] jMetal, Accessed 2016, http://jmetal.sourceforge.net/.

[79] Future Position X, Accessed 2017, http://www.fpx.se/.

[80] S. Yoo, M. Harman, Regression testing minimization, selection and prioritization: a
survey, Softw. Test. Verif. Reliab. 22 (2) (2012) 67–120. http://dx.doi.org/10.1002/
stv.430.

[81] S. Wang, S. Ali, A. Gotlieb, Cost-effective test suite minimization in product lines
using search techniques, J. Syst. Softw. 103 (2014) 370–391.

[82] S. Elbaum, A.G. Malishevsky, G. Rothermel, Test case prioritization: a family of
empirical studies. IEEE Trans. Softw. Eng. 28 (2) (2002) 159–182. http://dx.doi.
org/10.1109/32.988497.

[83] M. Harder, J. Mellen, M.D. Ernst, in: Improving test suites via operational abstraction, Proceedings of the 25th International Conference on Software Engineering, 2003, pp. 60–71.

[84] D. Di Nardo, N. Alshahwan, L. Briand, Y. Labiche, Coverage-based regression test case selection, minimization and prioritization: a case study on an industrial system, Softw. Test. Verif. Reliab. 25 (4) (2015) 371–396.

[85] R.C. Bryce, C.J. Colbourn, M.B. Cohen, in: A framework of greedy methods for constructing interaction test suites, Proceedings of the 27th International Conference on Software Engineering, 2005, pp. 146–155.

[86] R.C. Bryce, C.J. Colbourn, Test Prioritization for Pairwise Interaction Coverage, ACM, United States, 2005, pp. 1–7.

[87] S. McMaster, A. Memon, in: Call stack coverage for GUI test-suite reduction, 17th International Symposium on Software Reliability Engineering, 2006, pp. 33–44.

[88] J. Black, E. Melachrinoudis, D. Kaeli, in: Bi-criteria models for all-uses test suite reduction, Proceedings of the 26th International Conference on Software Engineering, 2004, pp. 106–115.

[89] M. Marr, A. Bertolino, Using spanning sets for coverage testing, IEEE Trans. Softw. Eng. 29 (11) (2003) 974–984. http://dx.doi.org/10.1109/tse.2003.1245299.

[90] D. Jeffrey, N. Gupta, in: Test suite reduction with selective redundancy, Software Maintenance (ICSM'05), 2005, pp. 549–558.

[91] Z. Li, M. Harman, R.M. Hierons, Search algorithms for regression test case prioritization, IEEE Trans. Softw. Eng. 33 (4) (2007) 225–237.

[92] T.A. Budd, Mutation Analysis of Program Test Data, ACM, United States, 1980.

[93] J.C. Munson, S.G. Elbaum, in: Code churn: a measure for estimating the impact of code change, Proceedings of 1998 International Conference on Software Maintenance, 1988, pp. 24–31.

[94] G. Rothermel, R.H. Untch, C. Chengyun, M.J. Harrold, Prioritizing test cases for regression testing. IEEE Trans. Softw. Eng. 27 (10) (2001) 929–948. http://dx.doi.org/10.1109/32.962562.

[95] S. Elbaum, A. Malishevsky, G. Rothermel, in: Incorporating varying test costs and fault severities into test case prioritization, Proceedings of the 23rd International Conference on Software Engineering. ICSE, 2001, pp. 329–338.

[96] H. Srikanth, L. Williams, J. Osborne, in: System test case prioritization of new and regression test cases, International Symposium on Empirical Software Engineering, 2005, p. 10.

[97] C. Hettiarachchi, H. Do, B. Choi, Risk-based test case prioritization using a fuzzy expert system, Inf. Softw. Technol. 69 (2016) 1–15. http://dx.doi.org/10.1016/j.infsof.2015.08.008.

[98] T. Ma, H. Zeng, X. Wang, in: Test case prioritization based on requirement correlations, 17th IEEE/ACIS International Conference on Software Engineering, Artificial Intelligence, Networking and Parallel/Distributed Computing (SNPD), 2016, pp. 419–424.

[99] R. Krishnamoorthi, S.A. Sahaaya Arul Mary, Factor oriented requirement coverage based system test case prioritization of new and regression test cases, Inf. Softw. Technol. 51 (4) (2009) 799–808. http://dx.doi.org/10.1016/j.infsof.2008.08.007.

[100] S. Tahvili, S. Mehrdad, L. Stig, A. Wasif, B. Markus, S. Daniel, in: Dynamic integration test selection based on test case dependencies, IEEE Ninth International Conference on Software Testing, Verification and Validation Workshops (ICSTW), 2016, pp. 277–286.

[101] B. Miranda, A. Bertolino, Scope-aided test prioritization, selection and minimization for software reuse, J. Syst. Softw. (2016). http://dx.doi.org/10.1016/j.jss.2016.06.058

[102] G. Fraser, F. Wotawa, Property relevant software testing with model-checkers, ACM SIGSOFT Softw. Eng. Notes 31 (6) (2006) 1–10.

[103] D. Pradhan, S. Wang, S. Ali, T. Yue, in: Search-based cost-effective test case selection within a time budget: an empirical study, Proceedings of the Genetic and Evolutionary Computation Conference, Denver, Colorado, USA, 2016, pp. 1085–1092.

[104] D. Pradhan, S. Wang, S. Ali, T. Yue, M. Liaaen, in: F. Wotawa, M. Nica, N. Kushik (Eds.), STIPI: using search to prioritize test cases based on multi-objectives derived from industrial practice, Testing Software and Systems: 28th IFIP WG 6.1 International Conference, ICTSS 2016, Graz, Austria, October 17–19, Proceedings, Springer International Publishing, Cham, 2016, pp. 172–190.

[105] S. Wang, S. Ali, T. Yue, Y. Li, M. Liaaen, in: A practical guide to select quality indicators for assessing pareto-based search algorithms in search-based software engineering, Proceedings of the 38th International Conference on Software Engineering, Austin, Texas, 2016, pp. 631–642.

[106] S. Wang, S. Ali, T. Yue, M. Liaaen, in: UPMOA: an improved search algorithm to support user-preference multi-objective optimization, IEEE 26th International Symposium on Software Reliability Engineering (ISSRE), 2015, pp. 393–404.

[107] S. Wang, S. Ali, A. Gotlieb, in: C. Le Goues, S. Yoo (Eds.), Random-weighted search-based multi-objective optimization revisited, Search-Based Software Engineering: 6th International Symposium, SSBSE 2014, Fortaleza, Brazil, August 26–29, 2014. Proceedings, Springer International Publishing, Cham, 2014, pp. 199–214.

[108] S. Yoo, M. Harman, in: Pareto efficient multi-objective test case selection, Proceedings of the 2007 International Symposium on Software Testing and Analysis, London, United Kingdom, 2007, pp. 140–150.

[109] S. Wang, S. Ali, A. Gotlieb, in: Minimizing test suites in software product lines using weight-based genetic algorithms, Proceeding of the Fifteenth Annual Conference on Genetic and Evolutionary Computation Conference, 2013, pp. 1493–1500.

[110] G. Fliedl, W. Mayerthaler, C. Winkler, C. Kop, H.C. Mayr, Enhancing Requirements Engineering by Natural Language Based Conceptual Predesign, in: F. Harashima, K. Ito, K. Tanie (Eds.), IEEE SMC '99, vol. 5, 1999, pp. 778–783.

[111] J.H. Mcdonald, Handbook of Biological Statistics, Sparky House Publishing, Baltimore, Maryland, 2009.

[112] D.J. Sheskin, Handbook of Parametric and Nonparametric Statistical Procedures, Chapman and Hall/CRC, United States, 2007.

[113] J. Latvakoski, H. Honka, Time simulation methods for testing protocol software embedded in communicating systems, in: C. Gyula, D. Margit, T. Katalin (Eds.), Testing of Communicating Systems, Springer, United States, 1999, pp. 379–394.

[114] R.H. Carver, K.-C. Tai, Replay and testing for concurrent programs, IEEE Soft. 8 (2) (1991) 66–74.

[115] S.N. Weiss, in: A formal framework for the study of concurrent program testing, Proceedings of the Second Workshop on Software Testing, Verification, and Analysis, 1988, pp. 106–113.

[116] W. Schutz, in: A test strategy for the distributed real-time system MARS, CompEuro'90. Proceedings of the 1990 IEEE International Conference on Computer Systems and Software Engineering, 1990, pp. 20–27.

[117] M. Kim, S.T. Chanson, S. Yoo, Design for testability of protocols based on formal specifications, in: C. Ana, B. Stan (Eds.), Protocol Test Systems VIII, Springer, United States, 1996, pp. 252–264.

[118] M.Z. Iqbal, A. Arcuri, L. Briand, Environment modeling with UML/MARTE to support black-box system testing for real-time embedded systems: methodology and industrial case studies, in: International Conference on Model Driven Engineering Languages and Systems, 2010, pp. 286–300.

[119] M.Z. Iqbal, A. Arcuri, L. Briand, Environment modeling and simulation for automated testing of soft real-time embedded software, Softw. Syst. Model. 14 (1) (2015) 483–524.

[120] M.Z. Iqbal, A. Arcuri, L. Briand, in: Combining search-based and adaptive random testing strategies for environment model-based testing of real-time embedded systems, International Symposium on Search Based Software Engineering, 2012, pp. 136–151.

[121] M.Z. Iqbal, A. Arcuri, L. Briand, in: Empirical investigation of search algorithms for environment model-based testing of real-time embedded software, Proceedings of the 2012 International Symposium on Software Testing and Analysis, 2012, pp. 199–209.

[122] A. Arcuri, M.Z. Iqbal, L. Briand, in: Black-box system testing of real-time embedded systems using random and search-based testing, IFIP International Conference on Testing Software and Systems, 2010, pp. 95–110.

[123] K.G. Larsen, M. Mikucionis, B. Nielsen, in: Online testing of real-time systems using UPPAAL, International Workshop on Formal Approaches to Software Testing, 2004, pp. 79–94.

[124] K.G. Larsen, M. Mikucionis, B. Nielsen, A. Skou, in: Testing real-time embedded software using UPPAAL-TRON: an industrial case study, Proceedings of the 5th ACM International Conference on Embedded Software, 2005, pp. 299–306.

[125] A. David, K.G. Larsen, S. Li, M. Mikucionis, B. Nielsen, in: Testing real-time systems under uncertainty, International Symposium on Formal Methods for Components and Objects, 2010, pp. 352–371.

[126] A. David, K.G. Larsen, S. Li, B. Nielsen, in: Timed testing under partial observability, International Conference on Software Testing Verification and Validation ICST'09, 2009, pp. 61–70.

[127] V. Garousi, in: Traffic-aware stress testing of distributed real-time systems based on UML models in the presence of time uncertainty, 1st International Conference on Software Testing, Verification, and Validation, 2008, pp. 92–101.

[128] S. Ali, Y. Li, T. Yue, M. Zhang, in: An empirical evaluation of mutation and crossover operators for multi-objective uncertainty-wise test minimization, in 10th International Workshop on Search-Based Software Testing, 2017.

[129] M. Gerhold, M. Stoelinga, in: Model-based testing of probabilistic systems, International Conference on Fundamental Approaches to Software Engineering, 2016, pp. 251–268.

[130] M. Gerhold, M. Stoelinga, Model-Based Testing of Stochastic Systems With IOCO Theory, ACM, United States, 2016.

[131] P. Bishop, L. Cyra, in: Overcoming non-determinism in testing smart devices: a case study, International Conference on Computer Safety, Reliability, and Security, 2010, pp. 237–250.

[132] M. Woehrle, K. Lampka, L. Thiele, Conformance testing for cyber-physical systems, ACM Trans. Softw. Eng. Methodol. 11 (4) (2012) 84.

[133] M. Woehrle, K. Lampka, L. Thiele, in: Segmented state space traversal for conformance testing of cyber-physical systems, International Conference on Formal Modeling and Analysis of Timed Systems, 2011, pp. 193–208.

[134] T. Kuroiwa, Y. Aoyama, N. Kushiro, Testing environment for CPS by cooperating model checking with execution testing, Proc. Comput. Sci. 96 (2016) 1341–1350.

[135] A. Arrieta, G. Sagardui, L. Etxeberria, J. Zander, Automatic generation of test system instances for configurable cyber-physical systems, Softw. Q. J. (2016) 1–43.

[136] H. Abbas, B. Hoxha, G. Fainekos, K. Ueda, Robustness-guided temporal logic testing and verification for stochastic cyber-physical systems, in: 4th Annual International Conference on Cyber Technology in Automation, Control, and Intelligent Systems (IEEE-CYBER), 2014, pp. 1–6.

[137] B.K. Aichernig, H. Brandl, E. Jöbstl, W. Krenn, in: Model-based mutation testing of hybrid systems, International Symposium on Formal Methods for Components and Objects, 2009, pp. 228–249.

[138] H. Roehm, J. Oehlerking, M. Woehrle, M. Althoff, in: Reachset conformance testing of hybrid automata, Proceedings of the 19th International Conference on Hybrid Systems: Computation and Control, 2016, pp. 277–286.

[139] T. Dang, T. Nahhal, Coverage-guided test generation for continuous and hybrid systems, Form. Method. Syst. Des. 34 (2) (2009) 183–213.

[140] T. Dang, N. Shalev, in: State estimation and property-guided exploration for hybrid systems testing, IFIP International Conference on Testing Software and Systems, 2012, pp. 152–167.

[141] Y. Qin, C. Xu, P. Yu, J. Lu, SIT: sampling-based interactive testing for self-adaptive apps, J. Syst. Softw. 120 (2016) 70–88.

[142] A.J. Ramirez, A.C. Jensen, B.H. Cheng, D.B. Knoester, Automatically exploring how uncertainty impacts behavior of dynamically adaptive systems, in: Proceedings of the 2011 26th IEEE/ACM International Conference on Automated Software Engineering, 2011, pp. 568–571.

[143] E.M. Fredericks, A.J. Ramirez, B.H. Cheng, in: Validating code-level behavior of dynamic adaptive systems in the face of uncertainty, International Symposium on Search Based Software Engineering, 2013, pp. 81–95.

[144] E.M. Fredericks, B. DeVries, B.H. Cheng, in: Towards run-time adaptation of test cases for self-adaptive systems in the face of uncertainty, Proceedings of the 9th International Symposium on Software Engineering for Adaptive and Self-Managing Systems, 2014, pp. 17–26.

[145] M. Harman, in: Making the case for MORTO: multi objective regression test optimization, Proceedings of the 2011 IEEE Fourth International Conference on Software Testing, Verification and Validation Workshops, 2011, pp. 111–114.

[146] A. Arrieta, S. Wang, G. Sagardui, L. Etxeberria, in: Test case prioritization of configurable cyber-physical systems with weight-based search algorithms, Proceedings of the 2016 on Genetic and Evolutionary Computation Conference (GECCO), Search-Based Software Engineering (SBSE) Track, 2016, pp. 1053–1060.

[147] A. Arrieta, S. Wang, G. Sagardui, L. Etxeberria, in: Search-based test case selection of cyber-physical system product lines for simulation-based validation, Proceedings of the 20th International Systems and Software Product Line Conference (SPLC), 2016, pp. 297–306.

[148] M. Harman, A. Mansouri, Y. Zhang, Search based software engineering: trends, techniques and applications, ACM Comput. Surv. 45 (1) (2012) 1–61 (Article No. 11).

[149] D. Pradhan, S. Wang, S. Ali, T. Yue, M. Liaaen, in: CBGA-ES: a cluster-based genetic algorithm with elitist selection for supporting multi-objective test optimization, Proceedings of 10th IEEE International Conference on Software Testing, Verification and Validation (ICST 2017), 2017.

[150] Search-Based Software Engineering Repository, Accessed 2005, http://crestweb.cs.ucl.ac.uk/resources/sbse_repository/.

[151] K.R. Walcott, M.L. Soffa, G.M. Kapfhammer, R.S. Roos, in: Time aware test suite prioritization, Proceedings of the 2006 International Symposium on Software Testing and Analysis, Portland, Maine, USA, 2006, pp. 1–12.

[152] L. Zhang, S.-S. Hou, C. Guo, T. Xie, H. Mei, in: Time-aware test-case prioritization using integer linear programming, Proceedings of the Eighteenth International Symposium on Software Testing and Analysis, Chicago, IL, USA, 2009, pp. 213–224.

[153] D. Kahneman, A. Tversky, Prospect theory: an analysis of decision under risk, Econometrica (1979) 263–291.

[154] OMG, Structured Assurance Case Metamodel, http://www.omg.org/spec/SACM/, 2015.

[155] ISO/IEC/IEEE, ISO/IEC/IEEE 24765: Systems and Software Engineering—Vocabulary, http://www.iso.org/iso/catalogue_detail.htm?csnumber=50518, 2010.

[156] ISO/IEC/IEEE, ISO/IEC/IEEE systems and software engineering—architecture description, in: ISO/IEC/IEEE 42010:2011(E) (Revision of ISO/IEC 42010:2007 and IEEE Std 1471-2000), 2011, pp. 1–46. http://ieeexplore.ieee.org/xpl/articleDetails.jsp?arnumber=6129467.

[157] ISO/IEC/IEEE, ISO/IEC/IEEE 15288: Systems and Software Engineering—System Life Cycle Processes, http://www.iso.org/iso/home/store/catalogue_tc/catalogue_detail.htm?csnumber=63711, 2015.

[158] ISO/IEC, ISO/IEC 16085: Systems and Software Engineering—Life Cycle Processes—Risk Management, 2006.

[159] ISO/IEC, ISO/IEC 25010: Systems and Software Engineering—Systems and Software Quality Requirements and Evaluation (SQuaRE)—System and Software Quality Models, http://www.iso.org/iso/catalogue_detail.htm?csnumber=35733, 2011.

[160] ISO/IEC, ISO/IEC 15026-x: Systems and Software Engineering—Systems and Software Assurance, 2013.

[161] ISO/IEC, ISO/IEC 12207: Systems and Software Engineering—Software Life Cycle Processes, http://www.iso.org/iso/catalogue_detail?csnumber=43447, 2008.

[162] ISO/IEC, ISO/IEC Guide 98: Uncertainty of Measurement, 2009.

[163] ISO/IEC, ISO/IEC 10165-7: Information Technology—Open Systems Interconnection—Structure of Management Information: General Relationship Model, http://www.iso.org/iso/iso_catalogue/catalogue_tc/catalogue_detail.htm?csnumber=24160, 1997.

[164] IEC, IEC Guide 115: Application of Uncertainty of Measurement to Conformity Assessment Activities in the Electrotechnical Sector, https://webstore.iec.ch/publication/7524, 2007.

[165] IEC, IEC 31010: Risk Management—Risk Assessment Techniques, http://www.iso.org/iso/catalogue_detail?csnumber=51073, 2009.

[166] IEEE, IEEE standard for software quality assurance processes, in: IEEE Std 730-2014 (Revision of IEEE Std 730-2002), 2014, pp. 1–138. http://ieeexplore.ieee.org/stampPDF/getPDF.jsp?tp=&arnumber=6835311. http://ieeexplore.ieee.org/xpl/articleDetails.jsp?arnumber=6835311.

[167] IEEE, IEEE standard for a software quality metrics methodology, in: IEEE Std 1061-1998, 1998. p. i, http://ieeexplore.ieee.org/stampPDF/getPDF.jsp?tp=&arnumber=749159.

[168] IEEE, IEEE P2413: Standard for an Architectural Framework for the Internet of Things (IoT), http://standards.ieee.org/develop/project/2413.html.

[169] ISO, ISO 9000: Quality Management Systems—Fundamentals and Vocabulary, https://www.iso.org/standard/45481.html, 2015.

[170] ISO, ISO 31000: Risk Management, http://www.iso.org/iso/home/standards/iso31000.htm, 2009.

[171] ISO, ISO 3534-1: Statistics—Vocabulary and Symbols—Part 1: General Statistical Terms and Terms Used in Probability, http://www.iso.org/iso/catalogue_detail.htm?csnumber=40145, 2006.

[172] ISO, ISO 21748: Guidance for the Use of Repeatability, Reproducibility and Trueness Estimates in Measurement Uncertainty Estimation, http://www.iso.org/iso/catalogue_detail.htm?csnumber=46373, 2010.

[173] ISO/TR, ISO/TR 13587: Three Statistical Approaches for the Assessment and Interpretation of Measurement Uncertainty, 2012.

[174] ISO/TS, ISO/TS 17503: Statistical Methods of Uncertainty Evaluation—Guidance on Evaluation of Uncertainty Using Two-Factor Crossed Designs, http://www.iso.org/iso/catalogue_detail.htm?csnumber=59900, 2015.

[175] ISO, ISO 9241-x: Ergonomics of Human-System Interaction, 2010.

[176] JCGM, JCGM 200: International Vocabulary of Metrology—Basic and General Concepts and Associated Terms, 2012.

[177] ETSI, TR 102 422 Methods for Testing and Specification (MTS) IMS Network Integration Testing Infrastructure Testing Methodology (v1.1.1), http://www.etsi.org/deliver/etsi_tr/102400_102499/102422/01.01.01_60/tr_102422v010101p.pdf, 2005.

[178] ETSI, ETSI EG 203 120 Methods for Testing and Specification (MTS); Model-Based Testing (MBT); Methodology for Standardized Test Specification Development (v1.1.1), http://www.etsi.org/deliver/etsi_eg/203100_203199/203130/01.01.01_50/eg_203130v010101m.pdf, 2013.

[179] ETSI, ETSI TR 101 583 Methods for Testing and Specification (MTS); Security Testing; Basic Terminology (v1.1.1), http://www.etsi.org/deliver/etsi_tr/101500_101599/101583/01.01.01_60/tr_101583v010101p.pdf, 2015.

[180] ETSI, ES 201 873-x TTCN-3 Specifications, http://www.ttcn-3.org/index.php/downloads/standards.

[181] ETSI, ETSI ES 203 119-x Methods for Testing and Specification (MTS). The Test Description Language (TDL), http://www.etsi.org/technologies-clusters/technologies/test-description-language.

[182] ETSI, ETSI TR 102 840 Methods for Testing and Specifications (MTS). Model-Based Testing in Standardisation (v1.2.1), http://www.etsi.org/deliver/etsi_tr/102800_102899/102840/01.02.01_60/tr_102840v010201p.pdf, 2011.

[183] OASIS, Open Services for Lifecycle Collaboration Core Specifications Version 3.0, http://open-services.net/wiki/core/Specification-3.0/OSLCCoreSpecificationsV3/OSLCCoreSpecificationsV3/, 2013.

ABOUT THE AUTHORS

Shaukat Ali is currently a senior research scientist in Simula Research Laboratory, Norway. His research focuses on devising novel methods for Verification and Validation (V&V) of large scale highly connected software-based systems that are commonly referred to as Cyber-Physical Systems (CPSs). He has been involved in several basic research, research-based innovation, and innovation projects in the capacity of PI/Co-PI related to model-based testing (MBT), search-based software engineering, and model-based system engineering. He has rich experience of working in several countries including UK, Canada, Norway, and Pakistan. Shaukat has been on the program committees of several international conferences (e.g., MODELS, ICST, GECCO, and SSBSE) and also served as a reviewer for several software engineering journals (e.g., TSE, IST, SOSYM, JSS, and TEVC). He is also actively participating in defining international standards on software modeling in Object Management Group (OMG), notably a new standard on Uncertainty Modeling.

Hong Lu is currently a postdoctoral researcher in Simula Research Laboratory (Norway). She has obtained her Ph.D. degree from Beihang University, Beijing, China, 2016. Her research interests mainly focus on search-based software engineering, product line engineering, and empirical software engineering. She has been an external reviewer for several international journals, e.g., JSS and SoSyM and several international conferences such as MODELS, GECCO, QSIC, etc.

Shuai Wang is currently playing as a postdoctoral researcher in Simula Research Laboratory (Norway) after successfully obtaining Ph.D. degree with an honor from the University of Oslo in 2015. He holds broad research interests such as search-based software engineering, product line engineering, model-based testing, and empirical software engineering with more than 20 publications from well-recognized international journals (such as JSS, EMSE, and SOSYM) and high-reputed international conferences (such as ICSE, MODELS, ISSRE, ICST, SPLC). He also a recipient of an ACM distinguished paper award of MODELS 2013 (also best application track paper) and an outstanding reviewer award during 2015–2016 of the Information and Software Technology journal. He is a member of the ACM and IEEE computer society.

Tao Yue is a chief research scientist in Simula Research Laboratory, Oslo, Norway and she is also affiliated with the University of Oslo. She has received the Ph.D. degree in the Department of Systems and Computer Engineering at Carleton University, Ottawa, Canada in 2010. Before that, she was an aviation engineer and system engineer for seven years. She has nearly 20 years of experience of conducting industry-oriented research with a focus on model-based engineering (MBE) in various application domains such as Avionics, Maritime and Energy, Communications, Automated Industry, and Healthcare in several countries including Canada, Norway, and China. Her present research area is software engineering, with specific interests in requirements engineering, MBE, model-based testing, uncertainty-wise testing, uncertainty modeling, search-based software engineering, empirical software engineering, and product line engineering, with a particular focus on large-scale software systems such as Cyber-Physical Systems. She has been on the program and organization committees of several international conferences (e.g., MODELS, MODELSWARD, RE, SPLC). She is also on the editorial board of Empirical Software Engineering and also actively participating in defining international standards in Object Management Group (OMG), including Uncertainty Modeling, SysML, and UTP.

Man Zhang is a Ph.D. student in Simula Research Laboratory, Norway (2015 to present) and the Department of Informatics, University of Oslo, Norway. Previously, she obtained her Master degree in Computer Technology from Beihang University, Beijing, China (2012–2015). Her main research interests include model-based testing of cyber-physical systems, uncertainty modeling, model-based engineering, and requirements engineering.

CHAPTER THREE

Testing the Control-Flow, Data-Flow, and Time Aspects of Communication Systems: A Survey

Rachida Dssouli*, Ahmed Khoumsi†, Mounia Elqortobi*, Jamal Bentahar*

*CIISE, Concordia University, Montreal, QC, Canada
†Université de Sherbrooke, Sherbrooke, QC, Canada

Contents

1. Introduction to Communication Software System Testing   96
2. Basic Concepts of Testing   97
   2.1 Testing Objective   97
   2.2 Conformance Test Architectures   98
   2.3 Test Suite Composition   99
   2.4 Model-Based Testing for Communication Systems   100
3. Testing the Control and Data-Flow Aspects   102
   3.1 FSM-Based Testing: A Brief Summary   102
   3.2 EFSM Testing Challenges   104
   3.3 Test Generation Approaches Based on the Extended FSM Model 107
   3.4 Conclusion (EFSM-Based Test Generation Approaches)   112
4. Testing the Communication Aspect   115
   4.1 Introduction to the Communication Aspect   115
   4.2 An Overview of Existing Techniques   116
   4.3 Conclusion and Discussion   121
5. Testing the Time Aspect   121
   5.1 Introduction   121
   5.2 Models Used in Real-Time MBT   122
   5.3 Conformance Relations in MBT   129
   5.4 Real-Time MBT Approaches   130
   5.5 Specific Testing Environments   137
   5.6 Specific Testing Objectives   139
   5.7 Interesting Benchmark   142
   5.8 Conclusion   142
6. Discussion and Conclusion   142

Acknowledgment   143
References   143
About the Authors   154

Abstract

Communication software systems are considered as critical national infrastructures that support users, corporations, and governments. Their description, implementation, and testing are subject to verification by standardization bodies. Formal models have been investigated to establish the conformity of communication protocols to these systems' standards over the past four decades, including Finite State Machine (FSM) models for control aspects, Extended Finite State Machine (EFSM) models for both the control-flow and the data-flow aspects, and Timed Automata (TA) for modeling the time aspect. Testing is a labor-intensive activity within a system's life cycle. It is a part of software quality assurance and claims a large portion of the total development cycle. Testing automation will enhance both product quality and budget optimization. This chapter provides a survey of control-flow, data-flow, and time aspects of test sequence generation approaches for communication software systems.

1. INTRODUCTION TO COMMUNICATION SOFTWARE SYSTEM TESTING

Communication systems are considered to be critical infrastructures that support users, corporations, and governments. They are the backbone of networking and distributed computing systems. The basic component of a communication system is a communication protocol entity. Communication protocols contain all the rules that govern communication between components, applications, and systems [1,2]. The functionality and correctness of a protocol entity requires both a valid specification process and a conforming implementation. The correctness of a protocol implementation can be achieved using formal verification or exhaustive testing. Both these processes face hurdles, the first in terms of state explosion, and the second in terms of input domain, which is often very large or infinite. The problem of correctness is, for practical reasons and feasibility, shifted to another problem of conformance testing under a comprehensive coverage criteria [3]. This testing is part of software quality assurance.

Testing communication systems have benefited from several advances in many areas, among them the standardization of protocols' specifications, modeling and formalization, the availability of standards' specification languages such as SDL [4,5] and UML [6], better definitions of coverage criteria and their associated metrics, advances in test generation methods, test case descriptions languages such as TTCN [7], and tools development.

Testing is an error detection mechanism that applies a set of well-designed experiments to an implementation under test (IUT) to detect errors. A faulty implementation differs from its specification in terms of observable behavior or actual outputs that are different from the expected outputs [8,9]. The latest is extracted from a system's specification. Testing is also a cost function. The question in testing is *how to generate a test suite (TS) that has the maximum error detection power at minimum cost.*

Communication systems testing is influenced by three major factors: (1) the size of the input domain, which is often infinite; (2) the quality of the test suite (its effectiveness in detecting errors); and (3) the test architecture that allows the control and observation of the implementation behavior.

In general, the main challenge in testing is to generate the smallest set of test sequences with the highest fault detection capability. This ratio is always in line with the limitations of budget and scheduling. The questions *What to test, How to test, At which level to test, Which aspect to test, What is the quality of test sequence,* and *When to stop testing* are still unanswered for most types of software applications and complex systems. There is no one universal way, as this process will always depend on the software, the designers' knowledge, the complexity, the budget and time constraints, and other factors.

This chapter is composed of five sections. Section 1 is an introduction to testing communication software systems. Section 2 introduces some basic concepts. Section 3 provides a review of the testing of control-flow and data-flow aspects, and Section 4 reviews the testing of the communication aspect. Time aspect testing is reviewed in Section 5. Section 6 discusses and concludes this chapter, and is followed by the references.

In the following section we introduce some basic concepts that will facilitate the reading of this survey.

2. BASIC CONCEPTS OF TESTING

2.1 Testing Objective

Each testing activity is designed to fulfill a testing objective that corresponds to a particular type of testing. Testing can be passive that is observing the behavior of the system, or active which means the behavior is stimulated with the injection of input sequences. There are several types of testing for communicating systems, such as functional testing, whose objective is to check the functional behavior of a system's implementation [8,9]. Functional testing is always applied. Integration testing is used to check the interfunctioning of a system's dependent communicating components during their composition or integration. Performance tests are used to analyze the system's nonfunctional requirements, such as response times, load testing, stress testing, CPU usage, and memory requirements. Load testing evaluates the behavior of a system under heavy loads. Stress testing is required to analyze the behavior of a system under extreme and unusual conditions. Conformance testing determines to what extent an implementation is conforming to its specification. It is often performed in black box testing. Interoperability testing ensures that implementations that were designed to work with each other are truly meeting their communication requirements. Interoperability can be assessed via active testing in which inputs are injected into communication channels, followed by observation of their output or reactions. Interoperability can also be assessed via passive testing, which is basically observing the system's behavior [10]. Regression testing is used to retest an implementation after a modification or change. This type of testing can be applied in a system's maintenance phase. The challenge is to determine the impact "zone" of the changes and then design additional tests, update the test battery, and select appropriate subsets of tests to apply to the modified implementation. In today's software industry, regression testing can be seen as a form of features testing. Real-time testing is a specific test that addresses the testing of a system's time aspect. The objective is to test the behavior of a system in relation to its real-time requirements. Section 5 is dedicated to this type of testing. In the remaining part of this survey, we only focus on active testing for conformance testing.

2.2 Conformance Test Architectures

While black box, gray box, and white box testing define the access level to an implementation in terms of details, it is test architecture that defines the access to the implementation within the context of a test environment [3]. Test architecture allows testers to observe the behavior and/or to inject input to the IUT. Its design is key to the testability of an implementation.

A test architecture is composed of an upper tester (UT), a lower tester (LT), test coordination procedure function, and point of control and observation of the IUT. An upper tester (UT) is used for control and observation at the upper layer of the IUT, and a lower tester (LT) is used for the control and observation of the IUT's lower layer. A test coordination procedure is required for the coordination between upper and lower testers in a distributed test environment, as it ensures the synchronization between testing entities. The UT and the LT have access to the IUT via access points known as points of control and observation (PCOs).

For communication systems, and in particular for the purpose of certification, four basic architectures are defined: (1) local test architecture, (2) coordinated test architecture, (3) distributed test architecture, and (4) remote test architecture. Test architectures have different error detection powers. Local test architecture has the best error detection power, as both upper and lower testers are located within the IUT's node and all PCOs are available for testing. Coordinated test architecture has both upper and lower testers, with the PCO between the IUT and the lower tester accessible over the network at a remote location. The upper and lower testers implement a coordination procedure that enhances the choreography of a test suites' execution. The coordinated test architecture has an error detection power that is less than, or at best, equal to that of the local test architecture. Distributed test architecture is similar to the coordinated test architecture minus the coordination procedure. It is reputed to have less error detection power than the distributed coordinated test architecture. The last one is remote test architecture, which has a remote lower tester but lacks access to the upper level tester. In this test architecture, the IUT may not be fully tested, and some actions that need to be triggered by the application layer cannot be stimulated. The remote test architecture has the least error detection power; however, it is used for testing by an external body or test certification center.

2.3 Test Suite Composition

A test case is a basic element that composes a test suite. It is composed of a preamble, a body, and a postamble. The preamble specifies how to reach the target from the initial state or a stable state. The body is the target (what to test). The postamble may be empty, as very often it embodies a verification sequence, like as a signature, for example, that brings the test case to a stable state such as an initial state. Test cases generated based on a specification have expected outputs.

Tc: <preamble, body, postamble, expected outputs>.

A test verdict, expressed in terms of pass, fail, or inconclusive, is associated to each test case. A pass verdict means that based on the test outputs, the IUT conforms to its specification. A fail verdict means that an invalid output has been detected, and therefore, the IUT is nonconforming to its specification. An inconclusive verdict means that the conformity or nonconformity is not established based on the test outcomes. Additional tests are required to turn an inconclusive verdict into a pass or fail verdict.

A test suite (TS) is a set of test cases. Sometimes the test cases are organized according to a partial order relation to facilitate their execution or for test optimization. A test suite is a test artifact associated to a specification, a test method, and a test coverage that express the quality of a test suite.

2.4 Model-Based Testing for Communication Systems

For decades, testing communication protocols have benefitted from the use of formal models in their specification and testing [11]. Models such as Finite State Machines (FSMs), Extended Finite State Machines (EFSMs), Timed Finite State Machines (TFSMs), Timed Extended Finite State Machines (TEFSMs), Communicating Finite State Machines (CFSMs), and Communicating Extended Finite State Machines (CEFSMs) are widely used [12,13]. These models differ in how and to what degree they express the aspects to test or in the combination of aspects that are required for a testing purpose. Very often, a test will focus on only one aspect. If more than one aspect is used, the model will express a specific combination of aspects. The most important aspects to test are the control, the data, the time, and the communication. Each aspect requires a different modeling technique. Testing these aspects in combination requires a richer, more complex model. Specification and modeling are a human task and an engineering activity; both need training and experience.

Test generation based on models, as depicted in Fig. 1, reveals that models can be extracted manually from informal specification or automatically from formal specification languages such as SDL [4,5], and UML state diagrams [6]. An extracted model can be used as a testing model. Depending on the aspect to test, the associated test generation technique requires either a fault model to detect [14], which is the case for FSM-based techniques that are used for the control aspect, or a coverage criterion to satisfy, which is the case for testing the data aspect. All test generation techniques require a test strategy that helps to optimize the traversal of the model or of a structure represented by more refined models, such as graphs and trees. Test generation algorithms will generate *abstract test cases* that will compose a test suite,

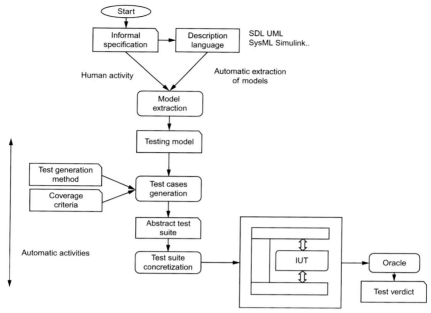

Fig. 1 Test sequences generation based on models.

taking into account the target coverage, test method, test strategy, and test objectives.

A *concrete test suite* is generated from an abstract test suite by adding variables and parameter values, as well as interface-related information. Only concrete test suites can be applied to an IUT.

The extracted model can be very abstract, such as a Labeled Transition System (LTS) or an FSM, which have limited expressiveness and can only model the control aspect of a communication protocol entity [13]. There are richer models, such as EFSM, which can express the data aspect, TFSM that expresses the time aspect that is a specific variable, CFSM that expresses the communication aspect between communication entities modeled as FSM, CEFSM that expresses the communication data and control aspects, and CTEFSM, which expresses all the aspects. The richer the model, the more challenging and complex the test generation technique will need to be.

Testing based on models is in fact trying to establish a relationship between the specification and the implementations derived from it. When extracting models from the specification and deriving test suites based on models, we establish a relationship between the model specification and an assumed abstract model of the IUT. We should be careful not to extrapolate those results to the entire IUT. For example, if we test the control aspect, then only that aspect can be the subject of inference relations under all the announced assumptions. The model for testing control aspect is limited; it cannot express data and time aspects. If the control aspect after testing is error free, this does not allow the tester to declare that the IUT is error free.

Conformance is a relation between a specification and an implementation [15]. In the case of an FSM specification that is deterministic and complete, the conformity relation is an equivalence relation between the FSM model specification and its corresponding FSM model implementation. In the case of a partially specified FSM model, the conformity relation between the FSM specification and its FSM implementation is called quasi-equivalence relation. For nondeterministic FSM (NFSM) models, several conformity relations have been proposed, including trace equivalence for completely specified NFSMs, trace quasi-equivalence for partially specified NFSMs, reduction for completely NFSMs, quasi-reduction for partially specified FSMs, as well as r–compatibility and nonseparability relations [16].

In the next section we will review testing of the control-flow and data-flow aspects of communication systems.

3. TESTING THE CONTROL AND DATA-FLOW ASPECTS

3.1 FSM-Based Testing: A Brief Summary

Most of the published research is related to formal model-based test sequence generation techniques [17]. These studies use one of the following models: FSM, NFSM, TFSM, CFSM, EFSM, or CEFSM, or a corresponding flow graph such as control-flow graph and data-flow graph. A test generation technique needs a coverage criterion to satisfy or a class of faults to detect. What to cover can only be found within the model's artifacts, such as states, transitions, nodes and edges for the control part, and data definitions and uses for the data flow.

The control aspect is normally represented by an FSM. Formally, a deterministic FSM is a 7-tuple $M = <I, O, S, S_0, D, \delta, \lambda>$, where I is the finite set of input symbols, O is the finite set of output symbols, S is the finite set of states, S_0 is the initial state, D is the specification domain, δ is the transfer function (also called the next-state function), and λ is the output function.

The transfer function and the output function together characterize the behavior of the FSM. For the specification domain $D = S \times I$, if the transfer function and the output function are defined for all state–input combinations of the FSM, then it is said to be completely specified (or completely defined).

Some assumptions are made about an FSM specification and its counterpart model implementation. The assumptions made for testing are as follows: completely or partially specified, deterministic or nondeterministic, minimized or reduced machine, and initially connected or strongly connected.

There are several well-known FSM-based test sequence generation methods for testing the control aspect known as *State Identification* techniques. We present a brief overview of six of them: the DS [18], UIO [19], UIOv [20], W [21], Wp [22], and the TT method [23]. Apart from the TT method, all of these methods are used as a signature in several EFSM and CEFSM test sequence generation methods. All state identification techniques have the same error detection power, but they differ in their size and complexity. A survey of the current FSM-based methods can be found in Lee *et al.* [24], and a more recent survey is given in Cavalli *et al.* [25].

The distinguishing sequence (DS) [18] method produces a signature that distinguishes among all states. It is an input sequence for a given FSM. If applied to each state in the FSM, it should produce a different output sequence for each state. The DS method is used under the assumption of deterministic, completely specified, and strongly connected FSMs. The disadvantage of this method is that a DS may not exist for a given FSM.

The Unique Input Output (UIO) [19] method distinguishes a state from all other states for a given FSM. It is known as a signature per state. This method can be applied if for each state of the specification, there is an input sequence such that the output produced by the machine, when it is initially in the given state, is different than that of all other states. The UIO has full fault coverage guarantee for an implementation that has the same number of states as the specification. If that is not the case, then UIOv, as proposed by Vuong *et al.*, has a complete fault coverage for such a case [20].

The W method is used to generate a set of small signatures or a finite set of input sequences that distinguish the behavior of any pair of states in the FSM [21]. The W method uses two sets called the W and the P sets. The W is a characterizing set that is a set of separating sequences. The P set is a transition cover set and provides a partial path to each transition in the FSM. P can be built as a transition cover tree using a normal breadth-first algorithm. A test sequence can be generated using a cross product between P and W with the assumption of a reset button.

The partial W or Wp method [21,22] is a generalization of the UIO method. It reduces the length of the test suite in comparison to the

W method. The test sequence generation is obtained in two phases. The first phase tests that every state defined in the specification exists in the implementation, using a Q set or state cover tree. The second phase is used to check that the remaining transitions $(P - Q)$ are correctly implemented. The Q set or state cover tree can be built using a normal breadth-first algorithm that provides a partial path to reach every state in the FSM. The test sequence for the first phase is obtained by "P.W" and the sequence for the second phase uses Wi sequences.

The TT method generates a test sequence called a *transition tour* [23]. This method is not part of the state identification methods, but it is often used because of its simplicity. For a given FSM, a transition tour is a sequence that takes the FSM from an initial state, traverses every transition at least once, and returns to the initial state. It is a variation of the Chinese Postman algorithm. The method guarantees the detection of all output faults, but there is no guarantee that all transfer faults can be detected. This method has a limited error detection power compared to identification-based methods. However, one advantage of this method is that the test sequences it obtains are usually shorter than the test sequences generated by the other methods.

The FSM-based techniques can be compared based on their error detection power, the length of their test sequences, the complexity of their algorithms, and implementation. As stated earlier, the TT method is not a state identification method. Its error detection power is weak, especially if we consider the transfer fault. However, TT is simple to apply, and thus it is often used for conformance testing of partially specified. The UIO, UIOv, DS, W, and Wp methods are state identification techniques that can guarantee a full fault coverage. UIO and UIOv produce shorter test sequences and are already supported by tools. UIO is widely used in many telecommunication industries. SUIO is a refinement of UIO that reduces the length of test sequences. Another variant of UIO, M-UIO, selects UIOs with a minimum length among the multiple UIO sequences of each state. In some cases, a DS may not exist. W and Wp are more general methods. They are always possible. W and Wp are also commonly offered in existing tools.

3.2 EFSM Testing Challenges

An EFSM is a Mealy machine with input and output that have parameters, internal variables, operations, and predicates that are defined over them.

An EFSM is formally defined as a 6-tuple $<S, s_0, I, O, T, V>$ where S is a nonempty set of states, $s_0 \in S$ is the initial state, I is a nonempty set of input interactions, O is a nonempty set of output interactions, T is a nonempty set of transitions, and V is the set variables.

A transition $t \in T$ is defined as a 5-tuple $t = <initial\ state, final\ state, input, predicate, block>$. The *initial state* and *final state* are the states in S representing the starting and the ending state, respectively, of a transition t. The *input* is either an input interaction from I or it is empty. *predicate* is a predicate expressed in terms of the variables in V, the parameters of the input interaction, and some constants. *block* is a set of assignment and output statements.

A system specified by an EFSM is tested for conformance by applying a sequence of inputs and verifying that the obtained sequence of observed outputs conforms to the expected output [24,25]. This type of testing is challenging, and brute force testing is not practical and very often infeasible [24]. EFSM test sequence generation is more complex than it is with FSM. An FSM model has no conditions along the paths that can affect their traversal; therefore, they are all feasible. However, the paths in EFSMs depend on the input, output, internal variables, operations, and predicates defined over them. Some predicates and condition expressions cannot be satisfied, resulting in nonexecutable paths. Test generation techniques that handle the control and data aspects are classified into three categories. The first category is where an EFSM is converted to an FSM and then identification-based methods such as UIO and W are applied [18–21]. This is possible if the variables' domains are finite and small enough. This technique will expand the set of states while accounting for the values of the variables and their combination. However, this technique is limited, as it generates the well-known problem of state explosion. The second technique is to use identification-based techniques for the control aspect and a data-flow graph to capture data dependencies [26,27]. The third technique uses a data-flow graph alone [28–31].

Most EFSM-based test case generation approaches are composed of the following components: (a) test sequence generation for a given coverage criterion, very often using one of the identification-based test case generation techniques; (b) test data selection to manage the data-flow aspect; and (c) path executability verification. The main challenge in EFSM-based testing is the path executability (feasibility) problem; it is very often undecidable and requires analysis techniques. The general testing-based EFSM process requires test case generation, test data generation and selection, executability analysis, and test results' analysis.

3.2.1 Test Data Generation and Selection Techniques

To test a protocol entity, both variables and parameters need values that must be selected from their domain definition. The selected data has the important role of stimulating the path and revealing any errors. The selected data should simultaneously satisfy all the predicates along the path for its feasibility (executability). The difficulty is that the input domain that combines all the variable and parameter domains is too large or infinite to consider a complete test suite from the data aspect. Test data generation is an undecidable problem [32].

Several techniques have been explored for test data generation and selection. One of the most common techniques is exhaustive testing, which refers to using every input sequence from the combinatorial set of all variable and parameter domains. Exhaustive testing can only be considered for very small EFSMs with tractable input domain size. To cope with large input domains, partition testing is preferred, as it consists of dividing the input domain into several equivalence classes from which only one test data is chosen [8,9]. The challenge is to define the equivalence relation that can best meet the requirements. In addition to exhaustive and partition testing, another technique used for software testing is boundary testing, which tests boundaries' limits [8,9]. Test data generation and selection techniques can be grouped into the following categories: symbolic execution [33–36], random [37], mutation [38], linear regression to narrow intervals, and search-based techniques [36].

3.2.2 Path Executability

Path executability is crucial for test sequence generation for models that express the control flow and data flow [39,40]. The problem can be addressed using several techniques. The first uses an expansion mechanism that transforms the model into an equivalent one by adding transitions and paths. Such a mechanism is similar to predicate decomposition and state decomposition [41], graph splitting and linear programming [42], and Transition Executability Analysis (TEA) [43,44]. Loop/cycle analysis is another technique that can be used for control and data flow in conjunction with constraint satisfaction [45–48]. More recently, researchers have been investigating search-based algorithms [48–50], genetic algorithms [51], and multiobjective optimization [39,52,53] with evaluation metrics or fitness functions. This domain remains wide open for further research.

3.2.3 Data-Flow Coverage Criteria

Data-flow testing focuses on data dependencies within a program and on how values are associated with variables [30]. A program variable is defined and used. Definition (def) assigns a new value to a variable; in this case, the variable appears on the left-hand side of an assignment statement or an input statement. A variable is used in one of the following ways: (1) computational use (c-use) if the variable is used within an expression on the right-hand side of an assignment statement or in an output statement, or (2) predicate use (p-use) if the variable affects the control flow; in this case, it appears within a condition expression.

Based on the above definitions, Rapps and Weyuker defined several coverage criteria for software testing [30]. The objective of test suite is to satisfy a certain coverage criteria and to expect that, with an efficient test data selection, all the existing errors will be detected. The criteria are: (1) all-defs, which cover each definition at least once; (2) all-uses, which cover each def-use association at least once; (3) all-du-paths that exercise all paths from each definition of a variable to each of their uses; (4) all c-uses, meaning all the variable definitions must be tested with all of the successive computational uses; (5) all p-uses: all variable definitions must be tested with all successive predicate uses; (6) all c-uses/some p-uses: in addition to the all c-uses criterion, a variable definition must be tested with a predicate use in case there is no successive computational use; (7) all p-uses/some c-uses, this last criterion means that, in addition to the all p-uses criterion, a variable definition must be tested with a computational use in case there is no successive predicate use. Even more specialized coverage criteria have been defined for EFSM-based testing, including IO-df-chain, which is an input–output data flow used to generate the do-paths (the definition to observation paths) [28].

3.3 Test Generation Approaches Based on the Extended FSM Model

Much research has been done on testing based on the EFSM. The EFSM is a model that extends the FSM model with variables and predicates that appear within condition statements. Test sequence generation is more complex in EFSM and very often faces a state explosion problem that leads to incomplete coverage. The data selection that is needed is undecidable and path executability is very expensive.

The first comprehensive survey was published by Bourhfir et al. in 1997 [26]. In 2014, Yang et al. published a more recent and exhaustive survey entitled "EFSM-based Test Case Generation: Sequences, Data and Oracle" [39]. However, it does not address the testing of the communication and time aspects. In the following paragraphs we give an overview of the most well-known test generation approaches, in chronological order.

In 1987, Sarikaya *et al.* [31] proposed an approach based on EFSM model. They transformed the specification to obtain normal form transitions such that it can be modeled by the control and data flow graphs for functional testing. The obtained graphs are decomposed into simpler functions of the protocol that represent the subtours and data-flow functions. Their approach allows the design of tests from the obtained subgraphs, considering parameter variations of the input primitives of each data-flow function and determining the expected outputs. The approach covers control paths and data-flow functions. The approach is not automatic. Coverage criteria are not specified; executability problem is not addressed in this work.

In 1991, Ural and Yang [28] were among the first researchers to generate test sequences for both control and data-flow aspects. They transformed the EFSM into control graph and data-flow graph and generated test sequences using IO-df-chain criterion to generate the do-paths. They identified definition–uses of variables and of input–output interaction parameter. The idea is to establish associations (IO-df-chains) between inputs and outputs. Test sequences can then be selected to cover each association at least once. The approach may not cover all transitions in certain cases. Path executability was not addressed.

In 1992, Miller and Paul [27] transformed an EFSM into an FSM by representing data with addition transitions using an expansion mechanism. The FSM is then transformed into a data-flow graph to generate test paths and test sequences to cover all-def-observation criterion for both the control and data flow. This criterion only includes those paths where the value of the input parameter is propagated to the output parameter. This method addresses control-flow testing by adding a UIO sequence for every state. It addresses the path executability problem by extracting a conditional path for every do-path, and checks its satisfaction. If the conditional path is not satisfied, it varies the value of the influencing variables. This technique uses backtracking to change the execution order of the conditional path. It addresses efficiency issues and helps to ensure that the variables are accessible to the tester. The drawback is this technique generates all the paths before checking their executability and then discards the infeasible paths.

In 1993 and 1994, Chanson and Zhu [45,46] transformed an EFSM into a Transition Dependence Graph (TDG) to represent control and data dependency between transitions. Their technique uses a Cyclic Characterizing Sequence (CCS) to verify the states for the control aspect. CCS concatenates existing characterization sequence (CS) method such as UIO [18–22] and the sequence that brings the machine back to the state where CS is applied. To cover the data flow, they use TDG graph to generate test sequences that cover the all-du-path criterion [30]. The path executability problem is addressed by analyzing self-loop, symbolic execution, and by using Constraint Satisfaction Problem (CSP) techniques [40,48]. The path nonexecutability analysis is carried by adding self-loops to change the value of the influencing variables. However, not all nonexecutable paths can have self-loops that solve the constraints. This technique generates all paths before evaluating their executability. The technique uses static loop analysis and symbolic evaluation techniques to determine how many times the self-loop should be repeated so that test cases become executable. In case of a specification where the influencing variable is not updated inside a self-loop, the technique is not appropriate. In general, the technique cannot be used if the number of loop iterations is not known.

In 1994, Li *et al.* proposed and extended Unique Input Output sequence, E-UIO, for a class of EFSM called EFSM/pres [54]. The EFSM is restricted with Presburger Arithmetic, in which expressions on input parameters and variables' expressions use simple operators such as addition, subtraction, and boolean operators. The EFSM for this approach should be completely specified. If all states have an E-UIO, then all paths are executable. For the general case, a state may not have an E-UIO sequence for each of its transitions. The imposed restriction only solves the problem of class of EFSM that does not represent real communication systems.

In 1994, Koh and Liu proposed a method to derive test paths that check both control flow and data flow for a protocol entity given as an EFSM model [47]. Their technique identifies a set of paths from an EFSM using one of the data-flow coverage criteria; then it appends a state check sequences to some transitions in this set of paths to check the control flow. To measure the control flow, some selected transitions from the set of derived paths will have state identification sequences appended to them. This method uses a criterion called the effective domain for testing to select those transitions. This criterion is used to evaluate how effective a transition can be when tested in a given path in order to distinguish between effective domains. The technique uses state identification techniques to selected transitions to check the control.

In 1995, Huang *et al.* proposed an approach that generates executable test sequences for both the control and data aspects from an EFSM [55]. They developed a solution that constructs a TEA tree from an EFSM that is similar to a reachability tree. A TEA tree is rooted to a given state in the EFSM configuration. They used the all–do–path criterion (with input and output associations) to generate do-paths from the TEA tree. The tree can be expanded using a breadth-first search (BFS) to generate executable test sequences. This TEA tree solves the executable problem by reachability analysis. All techniques that use a reachability tree must deal with the state explosion problem.

In 1995, Ramalingom *et al.* proposed a unified method to generate executable test cases for both control-flow and data-flow aspects for a protocol that is specified as an EFSM [56]. They consider the executability problem during the test sequence derivation. To check the control aspect, they introduced a new state identification sequence called the Context Independent Unique Sequence (CIUS) that is a variant of UIOv [20]. For data-flow coverage they used the all–uses criterion [30] extended with the def–use–ob criterion to enhance the observability. The technique uses a two-phase breadth-first algorithm to generate a set of executable test sequences to cover the above-specified criteria. The technique only addresses the executability of the preambles and postambles, and not with the executability of the du-paths covering the data flow. This technique has the same drawback as Chanson's work [45].

In 1997, Bourhfir *et al.* proposed an EFSM-based test sequence generation method that generates executable test sequences [57,58]. A complete test sequence is obtained in five steps. First, the technique transforms an EFSM model into a data-flow graph. Second, it selects input values for the input parameter that affects the control flow. Third, executable sequences are generated using du-paths and removing any subpath inclusion, while appending the state identification sequence and postamble to each du-path [30]. The executability of each path is verified in the fourth step. They used cycle analysis, symbolic execution, and CSP techniques to solve the path executability problem. Fifth, relevant paths are added to cover any uncovered transitions. This technique verifies the executability of a path during its generation. It uses optimized IO–df-chain [28] criterion and multiple UIO with Wp as the state identification method.

In 2002, Hierons *et al.* proposed a technique that bypasses the problem of path executability [41]. They derived a Normal Form EFSM (NF-EFSM) from an SDL-EFSM specification and then expanded the NF-EFSM into an EEFSM or a Partial EFSM (PEFSM). The focus of their research is on how to bypass the executability problem. They used a path-splitting technique that decomposes the predicates and transition conditions such that conflicting transitions are located on different paths. They only generate executable paths from EEFSM or PEFSM. Their test generation technique covers the all-uses criterion [30]. They assume that an NF–EFSM is deterministic, strongly connected, minimized, and completely specified. The EEFSM is similar to a reachability tree where all paths are executable. The technique suffers from the state explosion problem.

In 2004, Duale and Uyar proposed a technique that uses conflict avoidance [42], similar to the Hierons technique [41]. It detects and eliminates conflicts in EFSM models using graph-splitting methods, symbolic execution, and linear programming. This technique produces an EFSM graph in which all paths are executable and any existing test generation technique can be applied. However, this technique can only be applied to a subclass of EFSM where all conditions and actions are linear. The linearity assumption limits the use of this technique for real communication systems. It also suffers from the state explosion problem.

In 2004, Petrenko *et al.* proposed an approach that tries to find an identification-based sequence, called a Configuration-Confirming Sequence (CCS), which has the ability to distinguish a given configuration from a subset of configurations [59]. The given configuration is in fact a sort of mutant that has been distinguished for subset of other configurations. In this case, the EFSM in its expected configuration produces an output sequence that is different from that of any other configuration in that given set, or at least in a maximal proper subset. It uses the state identification-based method to identify the state as well as the given configuration/mutant. The authors do not address the issue of how to obtain a set of configurations.

In 2008, Wong *et al.* proposed a technique based on the identification of "hot spots" or areas with a high number of noncovered nodes that should be prioritized during test generation [60,61]. This approach helps to increase the coverage ratio of nodes and edges. It uses all-nodes and all-edges criteria [30]. The objective of this test sequence generation is to cover the identified hot spots using a constraint solver to verify the executability of a sequence. This approach also tries to generate an alternate sequence for the identified hot spot. The method is incremental with the number of hot spots and their prioritization. There is no state identification method or signature appended to the obtained sequences. The method inherits the same limitation as all techniques that use CSP to solve nonexecutable paths.

Several studies have addressed the issues of data selection and the path executability problem [52,62–65]. In 2009, Derderian *et al.* proposed an approach to solve the executability problem [66]. This approach uses a fitness function to estimate how easy (or difficult) it is to find an input sequence to trigger a given path within an EFSM. The proposed fitness function could be used in a search-based approach to find a path with good fitness to achieve a given test objective, such as executing a particular transition. This method also provides an estimation of a path's executability. Their technique can be used with any EFSM test sequence generation method. In 2009, Kalaji *et al.* proposed an approach based on genetic algorithms (GA) that guide the search for executable transition paths that are easy to trigger [63]. Their approach requires that the path length should be determined in advance and that this must be the same for all paths. Soon after, in 2010 they addressed the problem of a transition with a predicate that references a counter variable that is often the cause to path nonexecutability [64]. They proposed an approach that bypasses the counter problem and used a genetic algorithm to generate test data for a specified path, in which the fitness function uses a combination of a branch distance function and the approach level. Few researchers used search-based techniques to generate test sequences from EFSM [65,67,68]. Wu *et al.* proposed the use of McCabe complexity for EFSM and heuristic search to solve the problem.

Table 1 presents a summary and comparison criteria of the above-mentioned EFSM-based methods.

3.4 Conclusion (EFSM-Based Test Generation Approaches)

Testing the control and data-flow aspects of communication systems is a challenging task. The proposed techniques must address two additional problems than those associated with FSM-based methods: the executability of transition paths and data generation and selection. Some researchers have proposed to simply transform an EFSM into an FSM by expanding the number of transitions and decomposing states to allow path executability. These approaches must face a state explosion problem or incomplete path coverage [41,42]. Other approaches transform an EFSM into a graph that represents the control and data flow and then use graph-oriented paths and coverage criteria. These approaches do not use the state identification technique to check the ending state of transitions, which leads to weaker fault coverage but offers the satisfaction of graph-based coverage criteria, such as all-nodes, all-edges, all-uses, and all-def-clear paths [31]. Some approaches use loops

Table 1 EFSM-Based Test Sequences Generation Approaches [39]

Authors and Date	Model Transformation	Coverage Criteria	Signature	Data Selection	Path Executability Technique
Sarikaya 1987 [31]	EFSM to NF control graph and data-flow graph. Graph decomposition to functions	Control-flow paths Data-flow functions	—	—	—
Ural 1991, 2000 [28,29]	EFSM to control graph and data-flow graph	IO-df-chain	—	—	—
Miller 1992 [27]	EFSM to FSM (both control flow and data flow)	all-def-obs paths	UIO	—	Change the value of influencing variable, backtracking problem of efficiency
Chanson 1993, 1994 [45,46]	EFSM to transition dependency graphs	all-du paths	Any	—	Self-loop analysis, symbolic execution, CSP
Li 1994 [54]	EFSM/pres	all-transitions	E-UIO	—	—
Koh 1994 [47]	EFSM	du-path, data-flow coverage criteria	UIO	—	Transition effective domains, loop insertion
Huang 1995, 1998 [43,55]	EFSM	all-du path	UIO_E	Random	TEA
Ramalingom 1995 [56]	EFSM	def-use-ob paths	CIUS/ UIOv	—	—
Bourhfir 1997, 2001 [57,58]	EFSM to data-flow graph	IO-df-chain	M-UIO Wp	Random	Cycle analysis, symbolic execution, CLP-BNR technique

Continued

Table 1 EFSM-Based Test Sequences Generation Approaches [39]—cont'd

Authors and Date	Model Transformation	Coverage Criteria	Signature	Data Selection	Path Executability Technique
Hierons 2002 [41]	SDL–EFSM to NF–EFSM to EEFSM/PEFSM	all–uses	—	—	Path splitting, state decomposition, predicate decomposition, simplex algorithm
Duale 2004 [42]	EFSM to directed graph	all–uses	—	—	Graph splitting, simplex algorithm
Petrenko 2004 [59]	EFSM	—	CCS	—	—
Wong 2008, 2009 [60,61]	EFSM	all–nodes, all–edges, hot spots	—	Symbolic execution	Conflict detection, possibility to use CSP
Derderian 2009 [66]	EFSM	Any	—		Search–based algorithms, fitness function, evaluation metrics
Kalaji 2009, 2010 [63,64]	EFSM	all–transitions	—	—	Genetic algorithm, fitness functions
Wu 2012 [62]	EFSM	all–states, all–transitions	—	Symbolic execution	McCabe complexity for EFSM and heuristic search

—, means not given or not addressed.

and cycle analyses that can help produce only executable test sequences [46,57]. Others generate all paths and use a checking mechanism such as CPS to discard nonexecutable sequences; such approaches are expensive. More recently, researchers were focusing on new approaches for path executability and data generation [65,67,68]. Research has focused on the use of search-based techniques and genetic algorithms to solve the path executability problem [32,50,51,53,67,68]. For test data generation and selection, researchers have used techniques such as domain portioning, boundary analysis, linear programming, interval reduction, random generation, and symbolic execution.

4. TESTING THE COMMUNICATION ASPECT

4.1 Introduction to the Communication Aspect

There are many challenges to meet in concurrent processes testing [69,70]. Concurrent communicating machines have a distinct set of features, "such as communication, synchronization, and nondeterminism" [71], features that amplify the testing complexity [72]. In addition, the body of knowledge available for testing parallel communicating systems is not as mature as it is for sequential systems. The semantics of concurrent communicating processes needs to be defined in terms of interleaving actions or true concurrency, and the communication mode must be defined in terms of synchronous communication with blocking or in terms of asynchronous communication, a nonblocking mechanism. Parallel communicating machines are known for their nondeterministic behavior and their race condition. Nondeterminism is observed when we execute the same process with the same sequence of inputs and observe different behaviors, which means that different paths are being executed. In addition, the composition of deterministic processes may lead to nondeterministic behavior. Testing nondeterministic behavior requires testing multiple times with the same inputs, with the objective of testing the other branches of nondeterminism. This process leads to increased testing costs in the execution phase and often weaker fault coverage. Most of the techniques in testing concurrent processes use the composition of control graphs known as the Communication Finite State Machine (CFSM) or the Communicating Extended Finite State Machine (CEFSM). This composition, as well as the reachability tree obtained by cross products with interleaving semantics, leads to the well-known state explosion problem. It is in fact the same sort of reachability graph (RG) that is used for reachability analysis.

The approaches for testing CFSM/CEFSM can be classified into three types: (1) the composition of all single modules to obtain one composite model and the application of existing test sequence generation methods; (2) the generation of test sequences for single CFSM/CEFSMs, known as local test sequences, and then generating global test sequences by the composition of local test sequences to cover the set of communicating transitions; and (3) the transformation of each CFSM/CEFSM into control and data-flow graphs and the use of flow graph-based criteria to generate test sequences. In the following sections we review the case of CEFSMs that exchange messages through channels (asynchronous communication mode).

4.2 An Overview of Existing Techniques

In 1994, Luo *et al.* [71] proposed an approach that generates test sequences for concurrent and communicating protocol entities that are modeled as communicating nondeterministic finite state machines (CNFSMs). These CNFSMs are transformed into a single trace-equivalent NFSM using exhaustive reachability analysis. This approach uses an improved version of the Wp method [22] that generates test sequences for deterministic FSMs. The proposed approach has the same error detection power as Wp. However, it requires an assumption of a limited number of states in the implementations, such as that characterized by a slow environment. Most realistic communication systems are not in this category.

In 1996, Lee *et al.* proposed an approach to cover transitions in a CFSM using a guided random walk algorithm [72]. They use all-edges criterion. Transitions are covered with probability, which may challenge the complete coverage objective. They assume that they can observe the global states, but not the internal transitions. They defined the concepts of special observer such as semantics checker, safety observer, and progress observer to adapt their random walk algorithm. The technique requires a lot of internal information. It is a sort of gray box testing technique rather than conformance testing.

In 1997, Hierons addressed the testing a set of communicating FSM testing with a slow environment assumption. The proposed approach takes advantage of the machine's level of independence. If the machines are semi-independent, the problem of finding the shortest test set is reduced to the following relatively simple problems: testing the noncommunicating transitions, testing the nonfeedback communicating transitions, and testing the (weak) feedback transitions. He proposed an algorithm that computes the minimum test paths that cover all the communicating transitions of CFSMs. This algorithm only addresses the transitions that are triggered by an internal event. The other transitions are noncommunicating transitions (by default). It creates a graph with two machines to represent the relationships between noncommunicating transitions and communicating transitions and vice versa. These relationships lead to the creation of cycles. The objective is to find the shortest test paths that are then reduced to find a set of circuits with the minimum total cost. This is similar to a variant of the saving algorithm for a vehicle routing problem (VRP).

In 1998, Bourhfir *et al.* proposed an approach that automatically and incrementally generates executable test sequences for CEFSM model [73]. The communication mode is asynchronous. The approach does not compute the product of all communicating machines. It only generates test sequences by incrementally computing a partial product for each CEFSM; this means taking into account only transitions which influence (or are influenced by) the considered CEFSM and generating test sequences for it. The partial product for a CFSM represents its behavior when composed with parts of the other CEFSMs, the communicating transitions. They generate test sequences using the Extended Finite state machine Test Generator EFTG tool that was published in Ref. [57]. The tool generates executable test cases for EFSM-specified protocols which cover both control and data flow. The control-flow criterion used is the UIO (Unique Input Output) sequence [19] and the data-flow criterion is the all-definition-uses criterion [30]. Their approach is incremental and suitable for testing large systems. The objective is not to cover all transitions in the cross product of all CEFSMs, but to cover all transitions in all CEFSMs and all global transitions as well as all data-flow paths in each partial product. State explosion problem is possible in this technique.

In 2000, Henniger and Ural [74] proposed a model called the extended message flow graph (EMFG) that represents control and data dependencies both within a process as well as across process boundaries. The EMFG was designed to generate test sequences using control-flow-oriented as well as data-flow-oriented test selection criteria for communicating processes. Their EMFG and its construction rules are in part based on the adaptation of some earlier work for systems of asynchronously communicating state machines to SDL specifications.

In 2001, Maloku and Frey-Pucko [75] proposed an extension to EFSM models based on a flow-graph testing method that uses association of inputs, variables, and outputs, introduced by Ural and William [29] to address the issue of testing communicating EFSMs extracted from SDL specifications [4,5]. They address the nonexecutability issue by using simulation after test sequence generation. Test selection is guided by these associations and by the all-uses coverage criterion. The disadvantage of this approach is the state explosion problem that grows exponentially with the number of processes.

In 2002, Boroday et al. proposed an approach that merges two different coverage concepts into a test derivation strategy for CEFSMs. Their approach incorporates the coverage of interesting portions of the specification, augmented with a user-defined fault model. They reused their earlier work on confirming configurations to distinguish one configuration from other configurations or mutants' subset. Their approach inherits the main drawback of the original technique; it is difficult to determine the set of suspicious configurations that are introduced by a user-defined fault model.

In 2002, Li and Wong proposed a test generation approach based on the branching coverage of a specification given by CEFSM models. They created a flow diagram and used dominator, weight, and priority concepts to derive paths from the flow diagram to satisfy the branching coverage criterion [76]. A branch can represent a full transition in the absence of data-related decisions; otherwise it represents a part of a transition. The branching coverage criterion represents both the transition and the predicate coverage. The intention is to obtain a sufficient coverage with the smallest possible set of tests. A constraint-solving technique is needed to handle path executability. Their approach does not show how to identify the communicating transitions.

In 2003, Ambrosio [77] proposed an approach that generates test cases for each FSM in isolation, then creates a communication FSM that is composed of all the communication transitions among all the FSMs, which is a sort of composition of abstracted FSMs to their communicating transitions, and thereby generates test sequences for the communicating machine to obtain global test sequences [77]. Test sequences are generated using ConData tool that implements the transition-tour method with depth-first search [23]. The test sequences cover all paths with one-loop criterion. The approach is limited in its coverage and does not handle data and parameters.

In 2008, Wong and Lei [78] presented an approach that involves four automatic test sequences generation techniques for concurrent systems. Two of the approaches are based on hot spots and prioritization, and the two other approaches are based on topological sort. The objective is to generate a small set of test sequences that cover the all-nodes criterion in a reachability graph. They also use the same approaches to generate test sequences to cover all-edges criterion on a dual graph. The proposed approaches offer an incremental increase of the coverage using hot spots and prioritization. Achieving complete coverage with these approaches could be computationally expensive. The work is limited to CFSMs.

In 2012, Restrepo and Wong extended coverage criteria all-nodes and all-edges to address the issue of communication between SDL processes. In their previous work they covered the structure of an SDL process [61]. In this work, they extend that graph-oriented coverage of a single EFSM with new coverage criteria: n-step-message-transfer and sender–receiver-round-trip. Their test sequence generation uses backward tracking and forward validation. They do not incorporate a state identification technique to check the nodes, and data selection problem is not addressed.

In 2013, Yao *et al.* [79] proposed an approach that generates test sequences for parallel protocol components that communicate through shared variables for a Parallel Parameterized Extended Finite State Machine (PaP-EFSM) [79]. In their proposed approach, a dependency graph is created for all communicating machines using dependency relation based on Sync, which means a synchronization between the processes implementing the EFSM for a read operation required when a machine is reading an external variables. The dependency graph is assumed to be an acyclic graph that provides a hierarchy among communicating machines. Next, following a bottom-up approach, they generate a reachability graph for each machine. Starting from a lower-level machine, they generate nonexecutable test sequences for each machine using the Wp method, gathering a list of Sync events that will be used to generate test sequences. The access to external variables in synchronous communication mode is known as critical section and is protected by semaphore (lock and lock). Sync is the basis for establishing a partial order of inputs belonging to the test sequences involved in synchronization. Finally, following a top-down approach, executable test sequences are generated by first deriving local test sequences using the shortest path and forcing the execution of certain inputs to avoid nondeterminism. A set of definitions is stated in order to identify the context of a PaP-EFSM. This work is a formalization of testing specific CFSM in context that communicates with shared variables. It is conceptually similar to Bourhfir's work [73]. The executability problem is handled using reachability graph. However, it does face the state explosion problem, and data selection is not addressed.

Table 2 presents a summary and comparison criteria of the above-mentioned CFSM and CEFSM-based methods.

Table 2 CFSM/CEFSM-Based Test Sequences Generation Approaches [39]

Research Work	Model Transformation	Coverage Criteria	Signature	Test Generation Method
Luo et al. 1994 [71]	CNFSM to NFSM	Fault coverage Multiple faults	Generalized Wp	Wp
Lee et al. 1996 [72]	FSM to CFSM	all-transitions	—	Guided random walk observers
Hierons 1997 [80]	CFSM	Fault model	UIO	Heuristic VRP
Bourhfir et al. 1998 [73]	CEFSM to partial reachability graph	all def-uses all-transitions	UIO Wp	Random
Henniger and Ural 2000 [74]	EFSMs to flow graph	all-uses IO-df-chain	—	
Maloku and Frey-Pucko 2001 [75]	CEFSM to flow graph	all-uses	—	Combined flow graphs Checks path executability
Boroday et al. 2002 [81]	CEFSM	Specification and fault model	Confirming configuration	—
Li and Wong 2002 [76]	CEFSM to flow diagram	Branch coverage Transitions Predicates	—	Incremental test case generation based on coverage
Ambrosio 2003 [77]	FSM to abstract CFSM	All paths		Transition tour method Communication graph Test isolated FSM plus communication aspect
Wong and Lei 2008 [78]	CFSM to reachability graph	all-nodes all-edges	—	Hot-spot prioritization Topological sort

Table 2 CFSM/CEFSM-Based Test Sequences Generation Approaches [39]—cont'd

Research Work	Model Transformation	Coverage Criteria	Signature	Test Generation Method
Restrepo and Wong 2012 [82]	EFSMs	n–step–message–transfer sender–receiver–round-trip	—	
Yao *et al.* 2013 [79]	PaP-EFSM to reachability graph	def-use paths	Wp	Hot-spot prioritization Topological sort

—, means not given or not addressed.

4.3 Conclusion and Discussion

Testing the communication aspect is much more challenging than testing a single FSM/EFSM model. The most commonly used techniques are based on transforming the FSM/EFSM into a single model and then applying existing techniques. All of the solutions based on this composition face the state explosion problem [71,72]. Another possibility is to handle the communication transitions separately [77,82] and test each model in isolation, and to then compose test sequences. This type of solution faces a coverage problem, as the number of possible paths can be very large or even infinite. A third possibility is to use a partial reachability graph that optimizes the set of paths, but this suffers from the state explosion problem and may be challenging for large systems [73,79]. Lastly, some techniques reuse flow–graph–oriented techniques [75], which face the same hurdles as real communication systems. In addition, when a model expresses data and variables, the issue of path executability arises for those techniques that do not use a reachability graph. The issue of data generation and selection is common to all of these approaches.

5. TESTING THE TIME ASPECT

5.1 Introduction

So far, in this chapter, we have considered discrete event systems (DESs), i.e., systems whose behavior is defined by the possible orders of their interactions with the environment. Here we consider real-time DES (RTDES), whose behavior is also defined by the possible occurrence times of the interactions with the environment. Since an RTDES usually interacts with its environment through inputs and outputs, the environment has timing constraints for sending inputs to the RTDES, which in turn has timing constraints for sending outputs to its environment. (The terms "input" and "output" are utilized here from the viewpoint of the RTDES.)

RTDESs are tested to determine if they conform to a given specification. In order to develop rigorous real-time model-based testing (MBT) methods (i.e., MBT methods for RTDES), several formalisms have been developed to formally specify various known testing concepts in the real-time context, such as specification, test purpose, and conformance relation. Those formalisms extend the formalisms used in the nonreal-time case by introducing rigorous and formal representations of time and timing constraints.

RTDESs are encountered in various domains where the nonrespecting of timing constraints can have undesirable and even fatal consequences. These domains include, for example, telephone switching systems (e.g., to establish an urgent call), patient monitoring systems, and air traffic control systems. At the formal level, there may be strong analogies between those systems, because their complex parts usually consist of communicating software components. Due to this analogy, several testing methods presented in this section are not only applicable to communication systems but also to other software-based systems, sometimes with special adaptations.

The rest of this section is organized as follows: Section 5.2 introduces some of the main formalisms used in real-time MBT. In Section 5.3, we indicate the most used conformance relations in real-time MBT for RTDES in Section 5.3. Section 5.4 is an overview of the main MBT approaches. In Section 5.5, we discuss real-time MBT for specific testing environments, such as distributed and embedded real-time systems. Some particular real-time testing objectives are introduced in Section 5.6. Section 5.7 presents a benchmark that could be utilized to evaluate real-time MBT methods. A brief conclusion of real-time MBT is given in Section 5.8.

Since we only consider the testing context, an RTDES will be called a Real-Time System Under Test (RTSUT), and its environment will be called the tester.

5.2 Models Used in Real-Time MBT

5.2.1 Timed Automata

Timed automata (TA) [83–85] is the model that was the most successful at modeling an RTSUT and its specification, in the domain of communication systems as well as in other domains, such as software, control, and automotive systems. TA are based on the use of clocks, where a clock is a continuous variable that can be reset (i.e., set to 0) at any time and which evolves continuously so that its derivative with respect to time is equal to 1. Hence, the current value of a clock is the time that has elapsed since its last reset. Simply speaking, a timed automaton (TA) A is obtained from a finite state automaton (FSA) A and a set C of clocks by adding timing constraints to each transition of A as follows. Each transition Tr of A is associated to a reset Z which is a set of clocks, and a guard G which is a boolean expression depending on one or more clocks of C. Tr can be executed only if its guard G evaluates to true; the clocks in its reset Z are immediately set to 0 after the execution of Tr. Originally, each guard of a TA is a conjunction of timing constraints in the form "$c \sim k$," where c is a clock, k is a nonnegative integer, and $\sim$ is one of the operators $=, <, >, \leq, \geq$[83]. Such a TA type, called diagonal-free automata [86], presents several interesting properties that allow the development of several efficient analysis algorithms, such as those based on zone-graphs and the tool UPPAAL. We only consider diagonal-free automata, because they are the TA most commonly used in model-based techniques. This is in fact the basic form of diagonal-free automata; some variants and extensions will be mentioned later. Note that the nodes of the diagram representing a TA are called locations instead of states, because a state in a TA is defined by a location and by the current value of each clock.

As an example of TA, consider the two-state FSA executing alternately a and b. The FSA has two transitions: $Tr1$ that executes a and leads from state 1 to state 2, and $Tr2$ that executes event b and leads from state 2 to state 1. Consider the following three timing constraints specified using two clocks $c1$ and $c2$:

- The delay from a to b must be smaller than 2. This constraint is modeled by resetting $c1$ at the execution of $Tr1$ and stating $(c1 < 2)$ as a condition of the execution of $Tr2$.
- The delay from b to a must be smaller than 3. This constraint is modeled by resetting $c1$ at the execution of $Tr2$ and stating $(c1 < 3)$ as a condition of the execution of $Tr1$.
- The delay between two consecutive a must be smaller than 4. This constraint is modeled by resetting a clock $c2$ at the execution of $Tr1$ and stating $(c2 < 4)$ as a condition of the next execution of $Tr1$.

Therefore, $Tr1$ has a guard $G1 = (c1 < 3) \land (c2 < 4)$ and a reset $Z1 = \{c1, c2\}$, and $Tr2$ has a guard $G2 = (c2 < 2)$ and a reset $Z2 = \{c1\}$. The obtained TA is represented in Fig. 2, where the transitions are labeled in the form of "event; guard; reset."

The usual approach to analyze a system modeled by a TA is to construct an abstraction of the state space the TA, in the form of an FSA where time

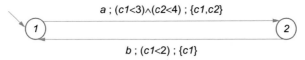

Fig. 2 Example of a TA.

passing is represented by a specific event. We present four abstractions that we believe are among the most important.

Region abstraction: The most precise abstraction of a TA is based on clock regions [83–85]. Intuitively, a clock region is defined by the integer part of current value of each clock and the order in which those integer parts will be incremented with time passing. Consider, for example, the following six situations of two clocks c_1 and c_2:

— situation 1: $c_1 = 1.25$ and $c_2 = 2.53$; situation 2: $c_1 = 1.33$ and $c_2 = 2.41$;
— situation 3: $c_1 = 1.2$ and $c_2 = 3.2$; situation 4: $c_1 = 1.37$ and $c_2 = 3.37$;
— situation 5: $c_1 = 1.5$ and $c_2 = 3.4$; situation 6: $c_1 = 1.7$ and $c_2 = 3.2$.

Situations 1 and 2 are in the same clock region, because for both situations the integer parts of c_1 and c_2 are 1 and 2, respectively, and with time passing the integer part of c_2 will be incremented before that of c_1 (indeed c_2 reaches the value 3 before c_1 reaches the value 2). Situations 3 and 4 are in the same clock region, as for both situations the integer parts of c_1 and c_2 are 1 and 3, respectively, and with time passing the integer parts of c_1 and c_2 will be incremented simultaneously. Situations 5 and 6 are in the same clock region, because for both situations the integer parts of c_1 and c_2 are 1 and 3, respectively, and with time passing the integer part of c_1 will be incremented before that of c_2.

The abstraction based on clock regions consists of generating an FSA called a region-graph, in which each state (called a region) is a combination of a location of the TA and a clock region. The transition between regions due to time passing is represented by a special event τ. The region-graph respects (all and only) the order and timing constraints of its corresponding TA. The fact that the region-graph is an FSA is a distinct advantage, as it implies that the FSA analysis techniques are applicable to these region-graphs. However, the drawback of region-graphs is that the number of regions is exponential in the number of clocks and polynomial in the maximum constant appearing in the guards for each clock. For example, the region-graph of the simple 1-clock TA of Fig. 3 has over 50 states [85].

One solution to reduce the exponential explosion of regions is to use a coarser and thus more compact representation of the state space. Zone-

Fig. 3 Simple one-clock TA whose region-graph has over 50 states [85].

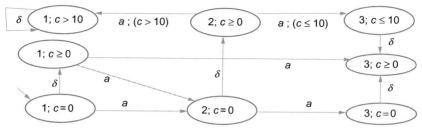

Fig. 4 Zone-graph of the TA of Fig. 3 [85].

graphs and SetExp automata have been developed for that purpose; we introduce them below.

Zone abstraction: The state space of a TA can be abstracted by an FSA called a zone-graph, whose states are called zones [85,87–92]. Each zone is defined by a location of the TA and a set of clock valuations. The latter set is specified as a conjunction of clock constraints, with each clock constraint in the form "$ci \sim k$" or "$ci - cj \sim k$," where ci and cj are clocks, k is a nonnegative integer, and $\sim$ is one of the operators $=, <, >, \leq, \geq$. The key idea when constructing a zone-graph is to maintain all valuations that are reachable along a path. The zone-graph can be infinite, but there exists an even coarser abstraction to obtain a (coarser) finite zone-graph. For example, Fig. 4 represents the zone-graph obtained from the TA of Fig. 4 [85]. This zone-graph has 8 states (instead of the over 50 states of the region-graph), where δ represents time passing.

SetExp abstraction: A TA can be abstracted by an FSA called a SetExp automaton, where time passing is modeled by Set and Exp events [93–95]. An event Set(c, k) means setting a clock c to 0 and programming its expiration that must occur when c reaches the value k. The event Exp(c,k) represents the programmed expiration. For example, Fig. 5 represents the SetExp automaton obtained from the TA of Fig. 3 "$x \parallel y$" denotes the simultaneous execution of x and y.

Tick abstraction: Instead of using a continuous time, time is approximated by an integer variable that is initially equal to 0 and then incremented by 1 after the passing of each time unit (e.g., second or minute, depending on the

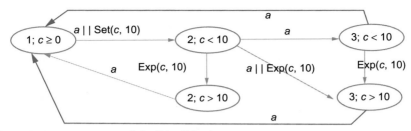

Fig. 5 SetExp automaton of the TA of Fig. 3.

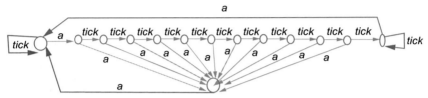

Fig. 6 Tick-automaton of the TA of Fig. 3.

selected precision). The passing of one time unit is modeled by a tick event generated by a fictitious global clock, such that all the clocks of the TA are incremented simultaneously with the tick event. Tick abstraction of a TA consists of representing the TA by an FSA called a tick-automaton, where time passing is modeled by the tick event [96–99]. Like region-graphs, tick-automata have the inconvenience of generating a state space explosion, which is, however, lower than that generated with region-graphs. Another disadvantage of the tick abstraction is that it induces an inaccuracy of at most two time units on delays between two events. For example, Fig. 6 represents the tick-automaton obtained from the TA of Fig. 3.

The four abstraction methods and some of their properties are represented in Table 3.

5.2.2 Specific TA

Specific TA that are suitable for real–time test studies have been proposed; we introduce some of these below.

TIOA, TAIO: The authors of Refs. [100–104] proposed *Timed Input Output Automata* (TIOA) which are a combination of TA and Input Output Automata (IOA) [105,106]. The main distinction between TIOA and TA is that the events are categorized into inputs and outputs. Inputs are sent by the tester to the RTSUT, while outputs are sent by the RTSUT to the tester.

Table 3 Abstraction Methods for Timed Automata

Abstraction Method	Research Work	Accuracy	Discrete Time (D) or Continuous Time (C)	State Explosion
Regions	[83–85]	Maximal	C	High
Zones	[85,87–92]	Acceptable (in general)	C	Reasonable (in general)
SetExp	[93–95]	Acceptable (in general)	C	Reasonable (in general)
Tick	[96–99]	High	D	High

Another similar model called *TA with Inputs and Outputs* (TAIO) is proposed in Refs. [107,108]. Their categorization of events into inputs and outputs is motivated by the fact that when testing an RTSUT, the tester must decide which inputs to send to the RTSUT, and verify which outputs are received from the RTSUT. In other words, the tester controls the inputs and observes the outputs.

DTA, ERA: Nondeterministic systems are systems whose current state is not necessarily determined by observing their execution. In general, nondeterministic TA cannot be determinized, and so some authors have proposed specific TA that can be determinized systematically. For example, Ref. [109] proposes *Determinizable TA* (DTA) where the clock reset of a transition depends uniquely on its event. Therefore, two transitions with the same event necessarily reset the same clock(s). Moreover, the reset of a transition *Tr* is empty if *Tr* executes an internal (i.e., unobservable) event. Refs. [110,111] propose *Event Recording Automata* (ERA), which are specific DTA where the resets of transitions are singletons.

TFSM: More than merely distinguishing inputs and outputs, some authors consider that there is a causality relation between inputs and outputs. Each transition is labeled in the form *i/o*, which means that the input *i* causes the output *o*. In particular, *i/−* means that *i* causes no output, and *−/o* means that *o* is spontaneous (i.e., it is not the consequence of an input). For example, Ref. [112] proposes *Timed Finite State Machines* (TFSM), where each transition is labeled in the form *i/o* with a possible constraint on the delay from *i* to *o*. Ref. [113] extends the TFSM of Ref. [112] by allowing timing constraints to also be specified on inputs as follows: for a transition-labeled *i/o* executable from a state *q*, we can specify a constraint on the delay from reaching *q* to executing *i*. The TFSM timing constraints of Refs. [112,113] are specified in the form such that a delay must belong to an interval. Ref. [114] restricts the TFSM of Ref. [113] by using timing constraints on the outputs, such that a delay must be equal to a specific value instead of belonging to an interval. Those TFSMs are specific TA, because their timing constraints are only between consecutive events.

5.2.3 Variants and Extensions of TA

Here we introduce several variants and extensions of TA that have been developed in the context of real-time testing.

TA with urgent transitions: The TA guards specify when events *can* occur, but they do not allow to specify when events *must* occur, for example, we cannot specify a deadline for an event. Refs. [115–117] introduce the notion of urgent transition in the sense that a transition must occur before a deadline. Such urgency is realized by using invariants, where an invariant of a location ℓ is a timing constraint such that ℓ must be left before its invariant is violated by time passing. Refs. [118,119] propose a different approach to define urgency, where a transition Tr is made urgent by associating to it an interval of time. Hence, an urgent transition Tr has a timing constraint specifying when it *must* occur, in addition to a possible guard specifying when it *can* occur.

VDTA: Ref. [120] defines *Variable-Driven TA* (VDTA), in which an event is an input or an output that corresponds to assigning values to the continuous variables of an RTSUT. More precisely, an input is an action of the tester over a memory-variable of the RTSUT that is controllable by the tester, and an output is an action of the RTDES over a memory-variable of the RTDES that is observable by the tester. Moreover, all transitions in a VDTA are urgent in the sense that they are fired as soon as they are enabled.

TIOSA, TIOSTS: Ref. [121] proposes Input Output Symbolic Transition Systems (IOSTS), an extension of FSA in which transitions can be conditioned by variables and variables can be set to specific values (not necessarily 0). Also, an event σ of IOSTS corresponds to sending or receiving a specific set of parameters. Refs. [122,123] propose *Timed Input Output Symbolic Automata* (TIOSA), a combination of the DTA of Ref. [109] and the IOSTS of Ref. [121]. Ref. [124] proposes a variant of TIOSA called *Timed IOSTS* (TIOSTS). The main difference between TIOSA and TIOSTS is that the latter are not necessarily determinizable and thus they can have transitions with deadlines.

LPTA, UPTA: Refs. [125,126] propose *Linearly Priced TA* (LPTA) which extends TA by associating prices to locations and transitions. The cost of executing a transition *Tr* is the price of *Tr*, and the cost of delaying *d* time units in a location l is the product of *d·p*, where *p* is the price associated with l. The cost of a trace (i.e., the sequence of consecutive transitions starting from the initial state) is the accumulated sum of the costs of its delays and transitions. LPTA can be analyzed, for example, to determine the minimum cost of traces ending in a goal state. *Uniformly Priced TA* (UPTA), defined by Ref. [125], are specific LPTA where all locations have the same price.

TIOGA, TGA: Refs. [127,128] propose *Timed I/O Game Automata* (TIGOA) and *Timed Game Automata* (TGA). TIOGA and TGA are game theory adaptations of TIOA and TA, respectively.

DATA*: Refs. [129,130] propose *Durational Action TA* (DATA*) which supports urgency and deadlines and whose actions have specified durations.

TMA: Moore Automata [131] are particular Mealy machines whose outputs are determined solely by their current state, where Mealy machines are particular input–output automata FSA whose outputs are determined both by their current state and their current inputs. Ref. [132] has proposed *Timed Moore Automata* (TMA) that extend Moore Automata in several aspects, in addition to defining timing constraints.

5.2.4 Timed Petri Nets
Several temporal extensions of Petri nets have been proposed in the literature. Ref. [133] proposes *Time Petri Nets* (TPN) which are obtained from Petri nets by associating time intervals to transitions. There are many other extensions of Petri Nets (PN), such as p-time PN [134] and timed PN [135], but none of them has been as successful as TPN.

In addition to timing constraints, other extensions also associate priorities to transitions. Ref. [136] proposes *Prioritized TPN* (PrTPN) which extend TPN by associating priorities to transitions. Ref. [137] proposes *Deterministic Input Enabled and Output Urgent-PrTPN* (DIEOU-PrTPN), which are specific PrTPN that allow the generation of time-optimal test cases.

5.3 Conformance Relations in MBT
Several different conformance relations have been proposed for real-time testing, because there is no consensus on a specific conformance relation. The selection of a conformance relation depends on the application and the context. For example, Ref. [138] studies several conformance relations, such as "RTSUT *is always faster than* the specification" and "RTSUT *is at*

least as fast as the worst case of the specification." We introduce some of the most commonly used conformance relations below. Henceforth, Spec denotes a model of the specification.

Trace equivalence: Let a trace denote a finite sequence $(e_1, \tau_1)(e_2, \tau_2) \cdots (e_n, \tau_n)$ that models the executions of observable events e_1, e_2, ... e_n at time instants τ_1, τ_2, ..., τ_n, respectively, such that $\tau_1 < \tau_2 < \cdots < \tau_n$. The intuition of the *trace equivalence conformance relation* is that RTSUT conforms to a Spec *if* RTSUT can execute *all and only* the traces that are accepted by the specification. Trace equivalence has been used by several authors, such as Refs. [100–104]. We obtain a *trace inclusion* relation if we replace "all and only" by "only."

tioco: The *timed input–output conformance relation* (tioco) defined in Refs. [97–99] is a timed extension of ioco [139]. The intuition of tioco is that RTSUT conforms to Spec if for each observable behavior specified in the Spec, the possible outputs of RTSUT after this behavior are a subset of the possible outputs of Spec. In tioco, time delays are included in the set of observable outputs. This makes it possible to identify that an implementation producing an output too early or too late (or never, whereas it should) is nonconforming. tioco is the most used conformance relation for real-time testing, e.g., in Refs. [97–99,107,108,140].

Some adaptations of tioco: Several adaptations of tioco have been developed. Three examples are: *tvco*, *rtioco*, and *dtioco*. The *timed variable-change conformance relation* (tvco) defined in Refs. [120,141] is an adaptation of tioco to the context of VDTA. The relativized tioco (rtioco) defined in Ref. [142] is an adaptation of tioco wherein assumptions on the environment's behavior are explicitly taken into account. The distributed tioco (dtioco) defined in Ref. [143] is an adaptation of tioco to distributed systems, which consists of applying tioco locally at each site of the distributed RTSUT.

The conformance relations mentioned in this section are represented in Table 4.

5.4 Real-Time MBT Approaches

The usual approach in real-time MBT is based on modeling the specification of RTSUT by using TA or a variant of TA (Section 5.2) and then constructing an abstraction of the state space of the model. The main state space abstractions adopted are those introduced in Section 5.2.1: region abstraction, zone abstraction, SetExp abstraction, and tick abstraction. The following three subsections (Sections 5.4.1–5.4.3) give an overview of MBT

Table 4 Conformance Relations for Real-Time Testing

Conformance Relation	Research Work	Some Specificities
Trace equivalence	[100–104]	Inputs and outputs not distinguished
tioco	[97–99]	Most used
tvco	[120,141]	Input = read data Output = write data
rtioco	[142]	Explicit assumptions on the environment
dtioco	[143]	Distributed architectures

methods based on region, zone, and SetExp abstractions, respectively. Then Section 5.4.4 introduces MBT based on other abstractions and approaches.

5.4.1 MBT Based on Region Abstraction

Several authors use the TIOA model and the region abstraction as the basis of their real-time MBT methods. From a specification and possibly a test purpose modeled by a TIOA, they construct a region-graph RG that preserves the state reachability of the TIOA. A RG contains an event τ that corresponds to the transition between adjacent regions due to time passing. Such τ is not practical in real-time testing, because it does not represent a specific delay. For this reason, Refs. [100–104] sample RG into an FSA called grid automaton (GA), in which the time passing is modeled by a constant delay δ depending on the selected granularity of the sampling. We say that the RG is reduced into a GA. Test cases are then extracted from the GA. A trace equivalence is used as the conformance relation. Ref. [101] assumes that the specification and RTSUT have deterministic behavior, and that an upper bound on the number of regions is known. Ref. [102] extends Ref. [101] to nondeterministic behaviors, and Ref. [144] presents a software tool based on the framework of Ref. [102]. Ref. [103] improves Refs. [100–102] with a guidance of test by using the test purposes. Ref. [104] improves Refs. [100–103] by making the method scalable based on the fact that a small number of test cases are generated, even for large systems.

Refs. [120,141] use the VDTA model and adapt the approach of Refs. [100–104] to the context of VDTA. They construct a region-graph that is sampled into a grid automaton from which test cases are generated. They use tvco [97–99] as a conformance relation. Their method is applicable to data-flow critical systems with time constraints.

Refs. [129,130] use the DATA* model, adapting the region abstraction to the context of DATA*. They construct a timed refusal region-graph (TRRG), which is a kind of region-graph to which are added two specific types of refusals: permanent and temporary. The TRRG is then transformed into a canonical tester from which test cases are generated. The meta-modeling tool AToM3 [145] is used to automatically generate a visual modeling tool to process models in DATA*, in TRRG, and in the canonical tester. A graph grammar is also used to define the transformation that takes the DATA* as input and generates a TRRG as output, passing through the canonical tester.

Refs. [107,108] model the specification, the RTSUT, and test cases by TAIO, and describe the test purpose using Open TAIO (OTAIO), which extends TAIO by using observed clocks whose values can be tested. In TAIO and in OTAIO, urgency is modeled with invariants. Since TAIO can be nondeterministic, an approximate determinization based on regions construction is applied using a game approach [146]. tioco [97–99] is used as a conformance relation.

Ref. [147] uses the TFSM model of Ref. [113] and applies an abstraction similar to regions. As in Ref. [101], it is assumed that an upper bound on the number of regions is known. They use two fault models, in which the input time instances are integer and rational, respectively.

Table 5 represents the above-mentioned MBT methods based on the region abstraction with some of their specificities.

5.4.2 MBT Based on or Inspired From Zone Abstraction

Among the possible state space abstractions of TA (Section 5.2.1), the region-graphs present the advantage of providing the best controllability to the tester, in the sense that the time instants of inputs (i.e., events sent from the tester to the RTSUT) can be selected with the best precision. However, the drawback of region-graphs is that the number of regions is exponential in the number of clocks, and polynomial in the maximum constant appearing in the guards for each clock. This is why the coarser abstraction based on zones has obtained more success in real-time analysis, especially in real-time model checking and testing. The development of software tools such as UPPAAL [148] has contributed to the success of analysis methods based on zone abstraction. In the following paragraphs, we present examples of real-time MBT methods based on zone abstraction. These methods are implemented by adapting and extending UPPAAL, a tool

Table 5 MBT Methods Based on the Region Abstraction

Model	Conformance Relation	Research Work	Some Specificities
TIOA	Trace equivalence or trace inclusion	[101]	The specification is deterministic The number of regions is bounded
		[102]	Generalizes Ref. [110] for nondeterministic specifications
		[103]	Improves Refs. [109,110,112] by the use of test purposes
		[104]	Improves Refs. [109,111,112] by making the test method scalable for complex systems
VDTA	tvco	[120,141]	Adapt [EDKE 1998, SVD 2001, EDK 2002, ED 2003, Enno 2008] to VDTA Applicable to data-flow critical real-time systems
DATA*	Based on refusal sets and timed traces (no specific name)	[129,130]	The interactions with the environment have duration
TAIO	tioco	[107,108]	Based on game theory
TFSM	f-compatibility	[147]	The number of regions is bounded Two fault models are used

initially developed for real-time model checking and essentially based on zone abstraction [148].

Ref. [149] proposes a test generation method based on both regions and zones. They also propose a test architecture to execute the generated test cases.

Refs. [110,111] use the ERA model and develop a test generation method based on zone abstraction.

Refs. [150,151] use the TA model and propose T-UPPAAL (UPPAAL for Testing Real-time systems Online) as an extension of UPPAAL to online real-time testing. T-UPPAAL generates and executes test cases online, i.e., test cases are executed immediately after their generation.

Ref. [142] uses the TA model and studies offline and online testing. They discuss how to specify test objectives and derive test sequences for both, as well as how to apply these to the system under test and how to assign a verdict. They use rtioco as the conformance relation. Off-line testing is studied by using UPPAAL COVER [152,153], an adaptation of UPPAAL for offline testing. A disadvantage of offline testing is that the specification is analyzed completely, which implies a state explosion. Another drawback of offline testing is that the determinization of TA is undecidable, and hence, nondeterministic TA are not supported. Due to these limitations of offline testing, the authors of Ref. [142] also study online testing by using UPPAAL-TRON, an extension of UPPAAL for online testing. In fact, UPPAAL-TRON appears to correspond to the T-UPPAAL of Refs. [150,151]. Compared to offline testing, online test generation reduces the state explosion problem and is adaptive to the nondeterminism of the specification and the RTSUT. A disadvantage of online testing is that it requires very efficient test generation algorithms, because they are executed online. Another inconvenient of online testing is that the test runs are typically long, and consequently, the cause of a test failure may be difficult to diagnose.

Ref. [154] uses the model of TA model and the UPPAAL COVER tool [152,153] for offline test generation. They study the problem of updating test cases for RTSUT whose behaviors have changed, without regenerating all the test cases. Instead, they only generate test cases of the affected parts of the behavioral model, i.e., the parts corresponding to modified behaviors and new behaviors.

Refs. [127,128] use the TIGOA and TGA models, respectively. The UPPAAL-TIGA tool [155,156] is used as an adaptation of UPPAAL that solves a timed game. The objective of the game corresponds to the test purpose. The two players are the tester and the RTSUT: the tester executes the (controllable) inputs, and the RTSUT executes the (uncontrollable) outputs and the internal actions. UPPAAL-TIGA will find a strategy for the tester to fulfill the test purpose if and only if such a strategy exists. Ref. [157] extends Refs. [127,128] with unobservable events. Ref. [158] extends Ref. [157] for remote testing by taking into account the communication delays between the tester and the RTSUT.

Ref. [140] uses the TIOSTS model that is abstracted by a symbolic execution tree in which each state is defined by a location, the conditions on clocks, and conditions on variables. Test cases are extracted offline from paths of the symbolic execution tree. tioco is used as a conformance relation.

Ref. [137] uses the DIEOU-PrTPN model, from which they construct a strong state class graph (SSCG) inspired from the TA zones [159,160]. Their automatic test generation method is based on their SSCG and realized by extending the TINA toolbox [161]. The generated test cases have the shortest possible accumulated time to be executed.

Ref. [162] extends the work of [151] to real-time hybrid systems (HSs) that combine RTDES and continuous systems. Typically, a HS consists of a continuous plant that interacts with a discrete controller. Real-time HSs are tested by running two tools in parallel. The UPPAAL-TRON tool [151] is used for testing the timed discrete behavior, while the PHAVER tool [163] is used for testing the continuous behavior. Note that TA are particular HSs, in which the clocks are the only continuous variables.

Table 6 represents the above-mentioned MBT methods based on the zone abstraction with some of their specificities.

5.4.3 MBT Based on SetExp Abstraction

SetExp abstraction [93–95] is another solution to reduce the exponential state space explosion of the region abstraction (Section 5.2.1). Ref. [96] applies SetExp abstraction to a discrete version of TA, where timing constraints are defined by timers that take integer values and are all incremented simultaneously by 1 with the constant frequency tick event generated by a fictitious global clock. The state space explosion due to tick abstraction is avoided, but not the 2-time-unit inaccuracy. Ref. [96] proposes a test architecture and a test case generation method based on this discrete version of SetExp abstraction.

Ref. [165] improves Ref. [96] in several aspects, such as the following two points: (1) a continuous time is used, and hence, the inaccuracy due to discrete time is removed; and (2) the number of generated test cases is reduced by using a more symbolic representation of test cases. Ref. [166] improves Ref. [165] by removing the restrictive assumption that the set of possible values of clocks in a location ℓ depends on the path that has been executed to reach ℓ. Ref. [166] also proposes a simpler test architecture than that in Refs. [96,165].

Ref. [109] considers more specifically the model of DTA, to which they apply the SetExp abstraction. Next they adapt the nonreal-time test generation method called Test Generation with Verification technology (TGV) to their own context [167]. Refs. [122,123] use the TIOSA models, which are a symbolic extension of DTA, using variables and parameters. They adapt the SetExp abstraction to the context of TIOSA and then propose a test generation method. Their work generalizes that of Refs. [121,168,169] to the real-time case.

Table 6 MBT Methods Based on the Zone Abstraction

Model	Conformance Relation	Research Work	Some Specificities
IOTA (similar to TIOA)	Timed trace equivalence	[164]	Propose a test architecture Use test purposes
ERA	Based on the state space of the specification	[110,111]	
TA	rtioco	[150,151]	Online testing using T-UPPAAL
		[142]	Off-line testing using UPPAAL COVER Online testing using UPPAAL-TRON
		[154]	Off-line testing using UPPAAL COVER
TGA, TIOGA	tioco	[127,128]	Based on game theory
		[157]	Based on game theory Generalizes Refs. [136,137] with unobservable events
		[158]	Based on game theory Generalizes Refs. [136,137] with remote testing
TIOST		[140]	
DIEOU-PrTPN	Timed trace inclusion	[137]	Use the tool TINA
Hybrid automata	Hybrid extension of rtioco	[162]	Extends Ref. [157] to real-time hybrid systems Testing discrete behavior using UPPAAL-TRON Testing continuous behavior using PHAVER

Table 7 represents the above-mentioned MBT methods based on the SetExp abstraction with some of their specificities.

5.4.4 A Few Other MBT Methods

Refs. [97–99] study the online (or on-the-fly) test generation of non-deterministic and partially observable RTSUT. Their adopted conformance

Table 7 MBT Methods Based on the SetExp Abstraction

Model	Conformance Relation	Research Work	Some Specificities
TA with discrete time	Timed trace inclusion	[96]	Use a discrete model of time Propose a test architecture
TA	Adaptation of tioco	[165]	Adapt Ref. [105] to a continuous model of time Improve Ref. [105] by reducing the number of test cases
		[166]	Improve Ref. [105] by: — weakening the assumptions on clock constraints — proposing a simpler test architecture
DTA		[109]	The specification can be nondeterministic
TIOSA	Generalization of tioco	[122,123]	Symbolic approach using variables and parameters

relation is tioco. They propose two types of online testing: analog-clock tests that precisely measure dense time, and digital-clock tests that measure time with the tick event. Digital-clock tests are relevant because only finite-precision clocks are available in practice. The test is adaptive, i.e., the action of the tester depends on the observation history. Test generation algorithms based on digital-clock tests are implemented in a tool called the TTG.

Ref. [132] uses the TMA model to specify the real-time behavior of cooperating reactive software components. They develop two test case generation methods. The first is based on Kripke structures and inspired from the model checking approach of Ref. [170]. The second test method relies on SAT solving techniques and is handled using the miniSAT tool [171].

5.5 Specific Testing Environments

In the following, we discuss two particular environments in real-time testing: distributed testing and embedded testing.

5.5.1 Embedded Real-Time Testing

Refs. [124,172] use the model of TIOSTS to model the specification of an embedded RTSUT, from which they generate test cases. The "embedded" aspect is taken into account by considering several issues related to the platform and the execution environment, where a particular attention is given to interruptions. Test sequences are generated in C code, so that the code can be instrumented and executed in an actual real-time environment. They use FreeRTOS as the execution platform [173].

Testing and Test Control Notation (TTCN-3) is a language for specifying scenarios for testing the functional behavior of telecommunication systems not only as a functional behavior of a system but also to satisfy timing constraints [3]. Timers are defined in TTCN-3 to specify time-out, but they are also practical to specify real-time requirements. The authors of Ref. [174] propose a temporal extension of TTCN-3 based on notions like: the time of an event, the time between two events, and predicates such as "at," "before," "after," and "within," as well as their negations. Such extension is a language to specify scenarios for testing not only a functional behavior of a system but also the satisfaction of timing constraints. Although TTCN-3 was developed for telecommunications systems, its temporal extension in Ref. [174] is motivated in the context of automotive systems, because the two domains have in common the integration and interoperation of devices and systems based on standards.

Ref. [114] studies the testing of a component P that is embedded in a system S consisting of several components. They assume that only P may be faulty in S, and that the tester does not interact directly with P. Test cases are generated to test the conformance of P while it is embedded in S.

5.5.2 Distributed Real-Time Testing

In distributed testing, we consider the situation where the RTSUT is distributed in different, distant sites. Only the aspects introduced by distribution are considered. We first introduce several distributed test methods based on the SetExp abstraction [175–181]. GTTS denotes a global timed test sequence.

We consider a distributed test architecture to test a distributed RTSUT. Given a GTTS that must be executed in a distributed testing architecture, the objective is to generate the corresponding local timed test sequences (LTTS$_i$) that must be executed at the various sites $i=1,...,n$. In Ref. [175], the clocks of the sites are assumed to be perfectly synchronized. Ref. [176] extends Ref. [175] by strengthening the timing constraints of a GTTS to make it executable, if necessary. Ref. [177] generalizes Ref. [175] to the case where the clocks are synchronized with a bounded error. Communication delays between the tester and the RTSUT are neglected in Refs. [175–177].

Ref. [181] considers the case where a GTTS has no timing constraints and communication delays are not negligible. Their objective is to determine the conditions on communication delays such that the tester is able to guarantee the order constraints of inputs and to verify the order constraints of outputs.

We consider now a *centralized* test architecture to test a distributed RTSUT. Given a GTTS, Ref. [178] computes the sufficient conditions on communication delays between the tester and RTSUT so that the tester is able to guarantee the timing constraints of inputs and verify the timing constraints of outputs. Ref. [179] improves Ref. [178] by providing weaker conditions on communication delays. Ref. [180] improves Ref. [179] by providing proofs of their results and additional explanations.

Ref. [143] considers the conformance relation dtioco, which depends uniquely on local timed test sequences (LTTS) and not on a GTTS. In other words, conformance is based on local orders and delays between events on the same site, but not between events at different sites. Hence, each local tester can verify the conformance of its local execution without communicating with the other local testers.

Refs. [182,183] study the causality between events at different sites and establish a global order of events by using their timestamps. They consider several cases of clock synchronization at the different sites: (1) the local clocks are perfectly synchronized; (2) the local clocks are synchronized with an error bounded by a constant; and (3) the local clocks are synchronized with a variable error. Communications between the tester and the RTSUT are assumed to be synchronous in Ref. [182], whereas Ref. [183] also considers asynchronous communications.

5.6 Specific Testing Objectives
5.6.1 Stress Testing

Refs. [184,185] consider a distributed RTSUT modeled with UML in the form of sequence diagrams (SD) annotated with precise timing information. The system's topology is given in a specific modeling format. The authors propose a stress test methodology (STM), referred to as Time-Shifting STM (TSSTM), to automatically generate sequences of network interactions that will stress the network traffic under strenuous conditions. TSSTM requires that the timing information of messages in SDs is available and as precise as possible. Ref. [186] proposes an STM, referred to as the Wait-Notify STM (WNSTM), which is an adaptation of the TSSTM that is applicable to stress test systems in which the timing information of messages is unpredictable or imprecise.

There are a few other works on the systematic generation of stress and load test suites for software systems [187–190]. For example, Ref. [189] proposes a procedure to automatically generate stress test cases in multimedia systems consisting of servers and clients connected through a network. Timing and synchronization constraints are required for the communications between servers and clients. Petri nets with timing constraints are used to model the flow and concurrency control of multimedia systems, and temporal logic is used to model timing constraints. As another example, Ref. [187] proposes algorithms to generate class test cases for telecommunication systems modeled by Markov chains.

5.6.2 Probabilistic Testing

There has been work that includes probabilities in models of real-time systems, e.g., Refs. [112,191–193]. More particularly, testing systems that combine stochastic and timing information have been studied, e.g., in Refs. [112,193]. In Ref. [112], the TFSM model is used, where each transition between two states is defined by an input i and an output o and a timing constraint between i and o. The authors consider three types of constraints on the time δ between i and o: (1) δ is equal to a value; (2) δ belongs to an interval; and (3) δ is a random variable following a given probability distribution function. For each of the three types of constraints, several conformance relations have been defined, because there is no consensus on a conformance relation in real-time systems. For example, depending on the application and the context, we may prefer that the RTSUT be faster than or as fast as the specification. Ref. [193] studies test derivation for stochastic nondeterministic TFSMs.

Ref. [194] uses Stochastic Timed I/O Automata (STIOA) from which they develop techniques to determine the optimum times in which the tester must feed the inputs of a given test case in order to achieve maximum probability of finding an error.

5.6.3 Passive Testing

In passive testing, the tester observes the behavior of the system under test without imposing any input to it, e.g., Refs. [10,195–197]. Ref. [10] studies nonreal-time passive testing and illustrates their approach with a Simple Connection Protocol. Ref. [195] generalizes Ref. [10] with a more general grammar to express invariants and illustrate their approach with Wireless Application Protocol. Ref. [196] generalizes Ref. [195] with timing constraints, and then Ref. [197] generalizes Ref. [196] with probabilistic timing constraints. Ref. [197] develops a tool called PASTE (for PASsive TEsting tool) that helps in the automation of their passive testing approach. Passive test can be used in several applications, such as in network management to detect configuration problems, fault identification, or resource provisioning [196], or to study the feasibility of new features as classes of services, network security, and congestion control [196].

5.6.4 Test Time Estimation

Some studies take into account test execution time. For example, Ref. [198] studies the problem of worst-case execution time testing that consists in finding the data that causes the longest execution time. They consider multicore systems, while previous research assumes sequential code running on single-core platforms. GA can be used to solve the optimization problem which is to search the worst-case inputs from the set of all possible inputs. They use search-based software engineering (SBSE) techniques based on genetic algorithms to find those inputs that will cause the program to take the longest execution time. The end-to-end time is measured by the Gem5 simulator [199].

As another example, Ref. [200] generates time–optimal test cases, i.e., test cases and suites that are guaranteed to take the least possible time to execute. Time optimality is inspired from algorithms from Refs. [126,201] that use Priced TA (PTA) and Uniformly PTA (UPTA), respectively. Outputs are urgent.

5.6.5 Modeling the Environment

Ref. [202] models the environment of the RTSUT using the UML/ MARTE standard, where MARTE is used to capture real-time properties. The environment models are used for code generation of the environment simulator, selecting test cases, and the generation of corresponding oracles. The selection of test cases is based on the models of the RTSUT and the environment, using heuristics to maximize chances of fault detection, like Adaptive Random Testing (ART) [203] and Search-Based Testing (SBT) [32]. Those heuristics have the property to be adjustable to available time and resources. The authors compare the results of the heuristic random testing (RT) [37], ART, and SBT, in a real-life example and also in artificial examples.

5.7 Interesting Benchmark

Ref. [204] presents a "real-world" example of a model used in the automotive industry for system test purposes. This model is public and suggested as a benchmark to evaluate test methods. Although the example corresponds to automotive industry, it can also be used to evaluate methods developed for other areas, such as communication systems. The authors propose to classify the test methods (implemented as test tools) by various criteria, e.g., by error detection capabilities of test cases, or by time to generate test data and number of resets. As a reference among the test tools to be evaluated, the authors propose a test generator method implemented in RT-Tester tool, which is an alternative to UPPAAL-TRON.

5.8 Conclusion

We note that the number of studies in real-time MBT has decreased significantly in the last 4 years. We think that the interest in this subject will be revived only if new fundamental ideas are developed, instead of continuing in the propositions of variants and adaptations of previous methods based on well-established approaches and tools, such as adaptations of the zone abstraction approach and the tool UPPAAL.

6. DISCUSSION AND CONCLUSION

Communication software testing is a challenging research domain. Up to now, there is no perfect solution to this testing problem. Progress has been made in control flow-based testing. Recent research has focused on variants of FSM model such as NFSMs and few refined conformity relations. For the testing of the combined control and data flow modeled by an EFSM, the problem is harder since it requires the satisfaction of a set of conditions along the path to be executed that requires path executability analysis, data generation, and selection techniques. The lasted are hot topics in particular with the use of search-based and genetic algorithms. This research domain is widely open and demands more research investigation. Testing the time aspect has been extensively studied and seems to reach a plateau. The problem of testing continuous data is in its early stage. It is expected that some new ideas in relation with continuous data selection that is present in avionics communication software can revive the interests in real-time testing. Another research area that is becoming a hot topic is run-time testing to ensure the quality of nonstop systems. Communication software requires updates without stopping them. It is also an issue addressed by standardization body.

In this chapter we reviewed test sequence generation that addresses the control–flow and data–flow aspects. We also dedicated an important part to review techniques dedicated to test the time aspect in communication software systems. We specifically addressed testing based on models such as EFSM, CFSM, CEFSM, and timed automata.

ACKNOWLEDGMENT

This work has been partly funded by NSERC CRD CRIAQ-CMC-CS Canada CRIAQ/AVIO-604.

REFERENCES

[1] G. von Bochmann, D. Rayner, C.H. West, Some notes on the history of protocol engineering, Comput. Netw. 54 (18) (2010) 3197–3209.

[2] G. von Bochmann, Protocol specification for OSI, Comput. Netw. ISDN Syst. 18 (3) (1989) 167–184.

[3] ISO/IEC, in: Information technology—open systems interconnection—conformance testing methodology and framework, International ISO/IEC Multipart Standard No. 9646, 1994–1997.

[4] ITU, Recommendation Z.100, ITU-T, Specification and Description Language, 1996.

[5] ITU-TS, Recommendation Z.100: Specification and Description Language, 2002.

[6] Object Management Group, Unified modeling Language Specification; Version 1.3. http://www.rational.com/uml, 1999.

[7] ETSI, ETSI Standard (ES) 201 873 V3.2.1: The Testing and Test Control Notation Version 3; Parts 1–8, in: European Telecommunications Standards Institute (ETSI), Sophia Antipolis, France, also published as ITU-T Recommendation series Z.140, 2007.

[8] B. Beizer, Software Testing Techniques (second ed.). International Thomson Computer Press, New York, NY, ISBN 978-0-442-20672-7, 1990.

[9] A. Mathur, Foundations of Software Testing, second ed., Pearson, Delhi, 2014.

[10] A. Cavalli, C. Gervy, S. Prokopenko, New approaches for passive testing using an extended finite state machine specification, Inf. Softw. Technol. 45 (12) (2003) 837–852.

[11] Z. Kohavi, Switching and Finite Automata Theory, McGraw-Hill, New York, 1978.

[12] I. Schieferdecker, Model based testing, IEEE Comput. Soc. IEEE Softw. 29 (1) (2012) 14–18.

[13] R. Lai, A survey of communication protocol testing, J. Syst. Softw. 62 (1) (2002) 21–46.

[14] G.v. Bochmann, A. Das, R. Dssouli, M. Dubuc, A. Ghedamsi, G. Luo, in: Fault model in testing, IFIP Transactions, Protocol Testing Systems. IV, Proceedings of the IFIP TC6 4th International Workshop on Protocol Test Systems, North-Holland, 1992, pp. 17–30.

[15] A. Petrenko, G.v. Bochmann, R. Dssouli, in: Conformance relations and test derivation, Proceedings of the IFIP Sixth International Workshop on Protocol Test Systems, Pau, France (Invited Paper), 1993.

[16] N. Yevtushenko, K. El-Fakih, A. Ermakov, in: On-the-fly construction of adaptive checking sequences for testing deterministic implementations of nondeterministic specifications, ICTSS, 2016, pp. 139–152.

[17] A.T. Dahbura, K.K. Sabnani, M.U. Uyar, Formal methods for generating protocol conformance test sequences, Proc. IEEE 78 (8) (1990) 1317–1325.

[18] G. Gonenc, A method for the design of fault detection experiments, IEEE Trans. Comput. 19 (1970) 551–558.

[19] K. Sabnani, A. Dahbura, A protocol test generation procedure, Comput. Netw. ISDN Syst. 15 (1988) 285–297.

[20] S.T. Vuong, W.L. Chan, M.R. Ito, in: The UIOv-method for protocol test sequence generation, Proceedings of the IFIP International Workshop on Protocol Test Systems, Berlin, Germany, 1989, pp. 161–175.

[21] T. S. Chow, "Testing software design modelled by finite state machine", IEEE Trans. Softw. Eng., Vol.4, No 3, May (1978), 178–187.

[22] S. Fujiwara, G. V. Bochmann, F. Khendec, M. Amalou, A. Ghedamsi, "Test selection based on finite state models", IEEE Trans. Softw. Eng., June (1991) 17 (6) 591–603.

[23] S. Naito, M. Tsunoyama, in: Fault detection for sequential machines by transition tours, Proceedings of the IEEE, Fault Tolerant Computing Conference, 1981, pp. 238–243.

[24] D. Lee, M. Yannakakis, Principles and methods of testing finite state machines: a survey, Proc. IEEE 84 (8) (1996) 1090–1123.

[25] R. Dorofeeva, K. El-Fakih, S. Maag, R. Cavalli, N. Yevtushenko, FSM-based conformance testing methods: a survey annotated with experimental evaluation, Inf. Softw. Technol. 52 (2010) 1286–1297.

[26] C. Bourhfir, R. Dssouli, E.M. Aboulhamid, in: Automatic Test Generation for EFSM-Based Systems, Technical Report IRO 1043, University of Montreal, 1996.

[27] R.E. Miller, S. Paul, in: Generating conformance test sequences for combined control and data flow of communication protocols, Proceedings of the IFIP Protocol Specifications, Testing and Verification (PSTV'92), 1992, pp. 13–27.

[28] H. Ural, B. Yang, A test sequence selection method for protocol testing, IEEE Trans. Commun. 39 (4) (1991) 514–523.

[29] H. Ural, A.W. Williams, in: Test generation by exposing control and data dependencies within system specifications in SDL, FORTE 1993, 1993, pp. 335–350.

[30] S. Rapps, E. Weyuker, Selecting software test data using data flow information, IEEE Trans. Softw. Eng. (1985) 367–375.

[31] B. Sarikaya, G.V. Bochmann, E. Cerny, A test design methodology for protocol testing, IEEE Trans. Softw. Eng. 13 (5) (1987) 518–531.

[32] P. McMinn, Search-based software test data generation: a survey, Softw. Test. Verif. Reliab. 14 (2) (2004) 105–156.

[33] P. Coward, Symbolic execution and testing, Inf. Softw. Technol. 33 (1) (1991) 53–64.

[34] L.A. Clarke, A system to generate test data and symbolically execute programs, IEEE Trans. Softw. Eng. 2 (3) (1976) 215–222.

[35] J.C. King, Symbolic execution and program testing, Commun. ACM 19 (7) (1976) 385–394.

[36] J. Zhang, C. Xu, X. Wang, in: Path-oriented test data generation using symbolic execution and constraint solving techniques, Proceedings of the 2nd International Conference on Software Engineering and Formal Methods, Beijing, China, 2004, pp. 242–250.

[37] G. Myers, The Art of Software Testing, Wiley, New York, 1979.

[38] M. Harman, Y. Jia, B. Langdon, in: Strong higher order mutation-based test data generation, Proceedings of the 8th European Software Engineering Conference and the ACM SIGSOFT Symposium on the Foundations of Software Engineering, Szeged, Hungary, 2011, pp. 212–222.

[39] R. Yang, Z. Chen, Z. Zhang, B. Xu, EFSM-based test case generation: sequence, data, and oracle, Int. J. Softw. Eng. Knowl. Eng. 25 (4) (2015) 633–667. (© World Scientific).

[40] D. Hedley, M.A. Hennell, in: The causes and effects of infeasible paths in computer programs, Proceedings of the 8th International Conference on Software Engineering, London, UK, 1985, pp. 259–266.

[41] R.M. Hierons, T.H. Kim, H. Ural, in: Expanding an extended finite state machine to aid testability, Proceedings of the 26th Annual International Computer Software and Applications, Oxford, UK, 2002, pp. 334–339.

[42] A.Y. Duale, M.U. Uyar, A method enabling feasible conformance test sequence generation for EFSM models, IEEE Trans. Comput. 53 (5) (2004) 614–627.

[43] C.M. Huang, Y.C. Lin, M.Y. Jang, in: An executable protocol test sequence generation method for EFSM-specified protocols, IFIP Transactions C: Communication Systems Protocol Test Systems, 1995, pp. 20–35.

[44] C.M. Huang, M.Y. Jang, Y.C. Lin, Executable EFSM-based data flow and control flow protocol test sequence generation using reachability analysis, J. Chin. Inst. Eng. 22 (5) (1999) 593–615.

[45] S.T. Chanson, J. Zhu, in: A unified approach to protocol test sequence generation, Proceedings of the IEEE Conference on Computer Communications (INFOCOM'93), San Francisco, USA, 1993, pp. 106–114.

[46] S.T. Chanson, J. Zhu, in: Automatic protocol test suite derivation, Proceedings of the 13th IEEE Networking for Global Communications (INFOCOM'94), Toronto, Canada, 1994, pp. 792–799.

[47] L. Koh, M. Liu, in: Test path selection based on effective domains, Proceedings of the International Conference on Network Protocols (ICNP'94), Boston, USA, 1994, pp. 64–71.

[48] J. Zhang, X. Wang, A constraint solver and its application to path feasibility analysis, Int. J. Softw. Eng. Knowl. Eng. 11 (2) (2001) 139–156.

[49] R. Zhao, M. Harman, Z. Li, in: Empirical study on the efficiency of search based test generation for EFSM models, Proceedings of the 3rd International Conference on Software Testing, Verification, and Validation Workshops, Paris, France, 2010, pp. 222–223.

[50] A.S. Kalaji, R.M. Hierons, S. Swift, An integrated search-based approach for automatic testing from extended finite state machine (EFSM) models, Inf. Softw. Technol. 53 (12) (2011) 1297–1318.

[51] R. Lefticaru, F. Ipate, in: Automatic state-based test generation using genetic algorithms, Proceedings of the 9th International Symposium on Symbolic and Numeric Algorithms for Scientific Computing, Timisoara, Romania, 2007, pp. 188–195.

[52] T. Yano, E. Martins, F.L. de Sousa, in: Generating feasible test paths from an executable model using a multi-objective approach, Proceedings of the 3rd International Conference on Software Testing, Verification, and Validation Workshops, Paris, France, 2010, pp. 236–239.

[53] T. Yano, E. Martins, F.L. de Sousa, in: MOST: a multi-objective search-based testing from EFSM, Proceedings of the 4th International Conference on Software Testing, Verification and Validation Workshops, IEEE Computer Society, Berlin, Germany, 2011, pp. 164–173.

[54] X. Li, T. Higashino, M. Higuchi, K. Taniguchi, Automatic generation of extended UIO sequences for communication protocols in an EFSM model, in: in: 7th International Workshop on Protocol Test Systems, Tokyo, Japan, November, 1994, pp. 225–240.

[55] C.M. Huang, M.S. Chiang, M.Y. Jang, UIOE: a protocol test sequence generation method using the transition executability analysis (TEA), Comput. Commun. 21 (16) (1998) 1462–1475.

[56] T. Ramalingom, A. Das, K. Thulasiraman, in: A unified test case generation method for the EFSM model using context independent unique sequences, Proceedings of the International Workshop on Protocol Test Systems (IWPTS'95), USA, 1995, pp. 289–305.

[57] C. Bourhfir, R. Dssouli, E. Aboulhamid, N. Rico, P.A. Aisenstadt, in: Automatic executable test case generation for extended finite state machine protocols, Proceedings of the the 10th International IFIP Workshop on Testing of Communicating Systems, Jeju Island, Korea, 1997, pp. 75–90.

[58] C. Bourhfir, E. Abdoulhamid, F. Khendek, R. Dssouli, Test cases selection from SDL specifications, Comput. Netw. 35 (6) (2001) 693–708.

[59] A. Petrenko, S. Boroday, R. Groz, Confirming configurations in EFSM testing, IEEE Trans. Softw. Eng. 30 (1) (2004) 29–42.

[60] W.E. Wong, A. Restrepo, Y. Qi, in: An EFSM-based test generation for validation of SDL specifications, Proceedings of the 3rd International Workshop on Automation of Software Test, Leipzig, Germany, 2008, pp. 25–32.

[61] W.E. Wong, A. Restrepo, B. Choi, Validation of SDL specifications using EFSM-based test generation, Inf. Softw. Technol. 51 (11) (2009) 1505–1519.

[62] T. Wu, J. Yan, J. Zhang, in: A path-oriented approach to generating executable test sequences for extended finite state machines, Proceedings of the 6th International Symposium on Theoretical Aspects of Software Engineering, Beijing, China, 2012, pp. 267–270.

[63] A.S. Kalaji, R.M. Hierons, S. Swift, in: Generating feasible transition paths for testing from an extended finite state machine (EFSM), Proceedings of the International Conference on Software Testing Verification and Validation, Denver, USA, 2009, pp. 230–239.

[64] A.S. Kalaji, R.M. Hierons, S. Swift, in: Generating feasible transition paths for testing from an extended finite state machine (EFSM) with the counter problem, Proceedings of the 3rd International Conference on Software Testing, Verification, and Validation Workshops, Paris, France, 2010, pp. 232–235.

[65] J. Zhang, R. Yang, Z.Y. Chen, Z.H. Zhao, B.W. Xu, in: Automated EFSM-based test case generation with scatter search, Proceedings of the International Conference on Software Engineering Workshop on Automated Software Test, Zurich, Switzerland, 2012, pp. 76–82.

[66] K. Derderian, R.M. Hierons, M. Harman, Q. Guo, Estimating the feasibility of transition paths in extended finite state machines, Autom. Softw. Eng. 17 (1) (2009) 33–56.

[67] F. Glover, in: A template for scatter search and path relinking, Proceedings of the Selected Papers from the Third European Conference on Artificial Evolution, London, UK, 1998, pp. 3–54.

[68] A.J. Nebro, F. Luna, E. Alba, B. Dorronsoro, J.J. Durillo, A. Beham, AbYSS: adapting scatter search to multi-objective optimization, IEEE Trans. Evol. Comput. 12 (4) (2008) 439–457.

[69] T. Hoare, Communicating Sequential Processes, Prentice-Hall, Inc. Upper Saddle River, NJ, 1985. ISBN: 0-13-153271-5

[70] R. Milner, Communication and Concurrency, Prentice Hall, Inc. Upper Saddle River, NJ, 1989. ISBN: 0131149849

[71] G.L. Luo, G.v. Bochmann, A. Petrenko, Test selection based on communicating nondeterministic finite-state machines using a generalized Wp-method, IEEE Trans. Softw. Eng. 20 (2) (1994) 149–161.

[72] D. Lee, K. Sabnani, D. Kristol, S. Paul, Conformance testing of protocols specified as communicating finite state machines—a guided random walk based approach, IEEE Trans. Commun. 44 (5) (1996) 631–640.

[73] C. Bourhfir, R. Dssouli, E. Aboulhamid, N. Rico, in: A guided incremental test case generation procedure for conformance testing for CEFSM specified protocols, IWTCS, 1998, pp. 275–290.

[74] O. Henniger, H. Ural, in: Test generation based on control and data dependencies within multi-process SDL specifications, SAM, 2000, pp. 189–202.

[75] N. Maloku, M. Frey-Pucko, in: SDL-based feasible test generation for communication protocols, Proceedings of the International Conference on Trends in Communication, Bratislava, Slovakia, July, 2001, pp. 536–539.

[76] J. Li, W. Wong, in: Automatic test generation from communicating extended finite state machine (CEFSM)-based models, Proceedings of the 5th IEEE International Symposium on Object-Oriented Real-Time Distributed Computing, ISORC, 2002, pp. 181–185.

[77] A.M. Ambrosio, in: Systematic test case generation for concurrent FSMs, International Conference on Dependable Systems and Networks (DSN), IEEE Computer Society, 2003, pp. A-46–A-48.

[78] W.E. Wong, Y. Lei, Reachability graph-based test sequence generation for concurrent programs, Int. J. Softw. Eng. Knowl. Eng. 18 (6) (2008) 803–822.

[79] J. Yao, Z. Wang, X. Yin, X. Shi, J. Wu, in: Reachability graph based hierarchical test generation for network protocols modeled as parallel finite state machines, 22nd International Conference on Computer Communication and Networks (ICCCN), 2013, pp. 1–9.

[80] R.M. Hierons, Testing from semi-independent communicating finite state machines with a slow environment, Softw. Eng. IEE Proc. 144 (5–6) (1997) 291–295.

[81] S. Boroday, A. Petrenko, R. Groz, Y.-M. Quemener, in: Test generation for CEFSM combining specification and fault coverage, TestCom, 2002, pp. 355–372.

[82] A. Restrepo, W.E. Wong, Validation of SDL-based architectural design models using communication-based coverage criteria, Inf. Softw. Technol. 54 (12) (2012) 1418–1431.

[83] R. Alur, D. Dill, A theory of timed automata, Theor. Comput. Sci. 126 (1994) 183–235.

[84] R. Alur, in: Timed automata, Proceedings of the 11th International Conference on Computer Aided Verification (CAV'99), LNCS 1633, Springer-Verlag, 1999, pp. 8–22.

[85] J. Bengtsson, W. Yi, in: Timed automata: semantics, algorithms and tools, Lectures on Concurrency and Petri Nets, LNCS 3098, 2003, pp. 87–124.

[86] B. Bérard, V. Diekert, P. Gastin, A. Petit, Characterization of the expressive power of silent transitions in timed automata, Fundam. Inform. 36 (1998) 145–182.

[87] D.L. Dill, in: Timing assumptions and verification of finite-state concurrent systems, Proceedings of the International Workshop on Automatic Verification Methods for Finite State Systems, LNCS 407, Springer-Verlag, 1989, pp. 197–212.

[88] M. Yannakakis, D. Lee, in: An efficient algorithm for minimizing real-time transition systems, Proceedings of the International Conference on Computer Aided Verification (CAV'93), LNCS 697, Springer-Verlag, 1993, pp. 210–224.

[89] T.G. Rokicki, Representing and modeling digital circuits, PhD thesisStanford University, 1993.

[90] W. Yi, P. Petterson, M. Daniels, in: Automatic verification of real-time communicating systems by constraint-solving, Proceedings of the IFIP Internatioanl Conference on Formal Description Techniques, 1994, pp. 223–238.

[91] T.A. Henzinger, X. Nicollin, J. Sifakis, S. Yovine, Symbolic model checking for real-time systems, J. Inf. Comput. 111 (2) (1994) 193–244.

[92] P. Pettersson, Modelling and verification of real-time systems using timed automata: theory and practice, PhD thesis, Uppsala University, 1999.

[93] L. Ouedraogo and A. Khoumsi, A new method for transforming timed automata, in Brazilian Symposium on Formal Methods (SBMF), Recife, Brazil, November 2004, In Electron. Notes Theor. Comput. Sci. 130 (2005) 101–128.

[94] L. Ouedraogo, A. Khoumsi, M. Nourelfath, in: Méthode de transformation d'automates temporisés avec invariants de localités, Conférence francophone de Modélisation et Simulation (MOSIM), Morocco, 2006(in french).

[95] L. Ouedraogo, A. Khoumsi, M. Nourelfath, SetExp: a method of transformation of timed automata, Real-Time Syst. 46 (2) (2010) 189–250.

[96] A. Khoumsi, M. Akalay, R. Dssouli, A. En-Nouaary, L. Granger, An approach for testing real time protocol entities, in: in: IFIP International Conference on Testing of Communicating Systems (TestCom'00), Testing of Communicating Systems, Series IFIP Advances in Information and Communication Technology, vol. 48, 2000, pp. 281–299.

[97] M. Krichen, S. Tripakis, in: Black-box conformance testing for real-time systems, 11th Internatioanl SPIN Workshop on Model Checking of Software, LNCS 2989: Springer, 2004, pp. 109–126.

[98] M. Krichen, S. Tripakis, in: F. Khendek, R. Dssouli (Eds.), An expressive and implementable formal framework for testing real time systems, IFIP International Conference on Testing of Communicating Systems (TestCom'05), LNCS 3502, Springer, 2005, pp. 209–225.

[99] M. Krichen, S. Tripakis, Conformance testing for real-time systems, Form. Methods Syst. Des. 34 (3) (2009) 238–304.

[100] A. En-Nouaary, R. Dssouli, F. Khendek, A. Elqortobi, in: Timed test cases generation based on state characterization technique, IEEE Real-Time Systems Symposium (RTSS'98), 1998, pp. 220–229.

[101] J. Springintveld, F. Vaadranger, P. Dargenio, Testing timed automata, Theor. Comput. Sci. 254 (1–2) (2001) 225–257.

[102] A. En-Nouaary, R. Dssouli, F. Khendek, Timed Wp-method: testing real-time systems, IEEE Trans. Softw. Eng. 28 (11) (2002) 1023–1038.

[103] A. En-Nouaary, R. Dssouli, in: D. Hogrefe, A. Wiles (Eds.), A guided method for testing timed input output automata, TestCom'03 Proceedings of the 15th IFIP International Conference on Testing of Communicating Systems, LNCS 2644, Springer, 2003, pp. 211–225.

[104] A. En-Nouaary, A scalable method for testing real-time systems, Softw. Qual. J. 16 (2008) 3–22.

[105] J. Tretmans, in: T. Margaria, B. SteKen (Eds.), Test generation with inputs, outputs, and quiescence, Proceedings of the 2nd Workshop on Tools and Algorithms for the Construction and Analysis of Systems (TACAS'96), Passau, Germany, LNCS 105, 1996, pp. 127–146.

[106] J. Tretmans, Test generation with inputs, outputs, and repetitive quiescence, Softw. Concepts Tools 17 (1996) 103–120.

[107] N. Bertrand, T. Jéron, A. Stainer, M. Krichen, in: Off-line test selection with test purposes for non-deterministic timed automata, Tools and Algorithms for the Construction and Analysis of Systems (TACAS'11), 2011, pp. 96–111.

[108] N. Bertrand, T. Jéron, A. Stainer, M. Krichen, Off-line test selection with test purposes for non-deterministic timed automata, Log. Meth. Comput. Sci. 8 (4:8) (2012) 1–33.

[109] A. Khoumsi, T. Jéron, H. Marchand, in: Test cases generation for nondeterministic real-time systems, Proceedings of the Formal Approaches to TEsting of Software (FATES'03), LNCS 2931, 2003, pp. 131–146.

[110] B. Nielsen, A. Skou, in: Automated test generation from timed automata, Tools and algorithms for the construction and analysis of systems (TACAS'01), LNCS 2031, Springer, 2001, pp. 131–146.

[111] B. Nielsen, A. Skou, Automated test generation from timed automata, Int. J. Softw. Tools Technol. Transf. 5 (1) (2003) 59–77.

[112] M. Merayo, M. Nuñez, I. Rodríguez, Formal testing from timed finite state machines, Comput. Netw. 52 (2) (2008) 432–460.

[113] M. Gromov, K. El-Fakih, N. Shabaldina, N. Yevtushenko, "Distinguishing non-deterministic timed finite state machines," in 11th Formal Methods for Open

Object-Based Distributed Systems and 29th Formal Techniques for Networked and Distributed Systems (FMOODS/FORTE'09). LNCS 5522, Springer (2009) 137–151.

[114] M. Gromov, A. Tvardovskii, N. Yevtushenko, in: Testing components of interacting timed finite state machines, IEEE East-West Design & Test Symposium (EWDTS'16), Yerevan, Armenia, October, 2016.

[115] S. Bornot, J. Sifakis, in: Relating time progress and deadlines in hybrid systems, International Workshop, HART'97, LNCS 1201, Springer, 1997, pp. 286–300.

[116] S. Bornot, J. Sifakis, S. Tripakis, in: Modeling urgency in timed systems, COMPOS'97, LNCS 1536, Springer, 1998, pp. 103–129.

[117] S. Bornot, J. Sifakis, An algebraic framework for urgency, Inform. Comput. 163 (1) (2000) 172–202.

[118] R. Barbuti, L. Tesci, in: Timed automata with urgent transitions, Proceedings of the International Workshop on Models for Time-Critical systems (MTCS'01), Aalborg, Denmark, August, 2001.

[119] R. Barbuti, L. Tesci, Timed automata with urgent transitions, Acta Inform. 40 (5) (2004) 317–347.

[120] O.N. Timo, A. Rollet, in: Conformance testing of variable driven automata, 8th IEEE Internatioanl Workshop on Factory Communication Systems (WFCS'10), Nancy, France, May, 2010.

[121] V. Rusu, L. du Bousquet, T. Jéron, in: An approach to symbolic test generation, International Conference on Integrating Formal Methods (IFM'00), LNCS 1945, 2000, pp. 338–357.

[122] A. Khoumsi, in: Complete test graph generation for symbolic real-time systems, Brazilian Symposium on Formal Methods (SBMF'04), Recife, Brazil, November, 2004.

[123] A. Khoumsi, On synthesizing test cases in symbolic real-time testing, J. Braz. Comput. Soc. 12 (2) (2006).

[124] W. Andrade, P. Machado, E. Alves, D. Almeida, in: Test case generation of embedded real-time systems with interruptions for FreeRTOS, Formal Methods: Foundations and Applications: Brazilian Symposium on Formal Methods (SBMF'09). LNCS 5902, Springer, 2009, pp. 54–69.

[125] G. Behrmann, A. Fehnker, T. Hune, K.G. Larsen, P. Pettersson, J. Romijn, F. Vaandrager, in: Minimum-cost reachability for priced timed automata, International Workshop on Hybrid Systems. Hybrid Systems: Computation and Control, HSCC, 2001, pp. 147–161.

[126] K.G. Larsen, G. Behrmann, E. Brinksma, A. Fehnker, T. Hune, P. Pettersson, J. Romijn, in: G. Berry, H. Comon, A. Finkel (Eds.), As cheap as possible: efficient cost-optimal reachability for priced timed automata, Proceedings of the International Conference on Computer Aided Verification (CAV'01), LNCS 2102, Springer-Verlag, 2001, pp. 493–505.

[127] A. David, K.G. Larsen, S. Li, B. Nielsen, Cooperative testing of timed systems, Electron. Notes Theor. Comput. Sci. 220 (1) (2008) 79–92.

[128] A. David, K.G. Larsen, S. Li, B. Nielsen, in: A game theoretic approach to real-time system testing, Design, Automation and Test in Europe (DATE'08), Munich, Germany, March, 2008.

[129] H. Hachichi, I. Kaitouni, K. Bouaroudj, D.-E. Saidouni, A graph transformation approach for testing timed systems, in: T. Skersys, R. Butleris, R. Butkiene (Eds.), Information and Software Technologies. ICIST 2012, Communications in Computer and Information Science, vol. 319, Springer, Berlin, Heidelberg, 2012.

[130] I. Kitouni, D.-E. Saidouni, K. Bouaroudj, in: Modeling and testing non-deterministic real-time systems, IFIP International Conference on Testing Software and Systems (ICTSS'12), Aalborg, Denmark, November, 2012.

[131] E.F. Moore, Gedanken-experiments on sequential machines, in: in: C.E. Shannon, J. McCarthy (Eds.), Automata Studies, Annals of Mathematical Studies, vol. 34, Princeton University Press, Princeton, NJ, 1956, pp. 129–153.

[132] H. Löding, J. Peleska, in: Timed moore automata: test data generation and model checking, IEEE International Conference on Software Testing, Verification and Validation (ICST'10), Paris, France, April, 2010.

[133] J.M. Merlin, J. Farber, Recoverability of communication protocols: implications of a theoretical study, IEEE Trans. Commun. 24 (9) (1976) 1036–1043.

[134] W. Khansa, Réseaux de Petri P-temporels: contribution à l'étude des systèmes à événements discrets, Université de Savoie, Annecy, 1997.

[135] C. Ramchadani, Analysis of asynchronous concurrent systems by timed Petri nets, PhD thesisMIT, Department of Electrical Engineering, Cambridge, MA, 1992.

[136] B. Berthomieu, F. Peres, F. Vernadat, in: E. Asarin, P. Bouyer (Eds.), Bridging the gap between timed automata and bounded time Petri nets, International Conference on Formal Modelling and Analysis of Timed Systems (FORMATS'06), LNCS 4202, Springer, 2006, pp. 82–97.

[137] N. Adjir, P. De Saqui-Sannes, K.M. Rahmouni, in: Testing real-time systems using TINA, IFIP International Conference on Testing of Communicating Systems and the Workshop on Formal Approaches to Testing of Software (TestCom/FATES'09), Eindhoven, The Netherlands, November, 2009.

[138] M. Nuñez, I. Rodríguez, in: W. Grieskamp, C. Weise (Eds.), Conformance testing relations for timed systems, Workshop on Formal Approaches to Testing of Software (FATES'05), LNCS, 3997, Springer, 2006, pp. 103–117.

[139] J. Tretmans, in: J.C.M. Baeten, S. Mauw (Eds.), Testing concurrent systems: a formal approach, 10th International Conference on Concurrency Theory (CONCUR'99), LNCS 1664, Springer-Verlag, 1999, pp. 46–65.

[140] B. Bannour, J.P. Escobedo, C. Gaston, P. Le Gall, in: Off-line test case generation for timed symbolic model-based conformance testing, IFIP International Conference on Testing Software and Systems (ICTSS'12), Aalborg, Denmark, November, 2012.

[141] O.N. Timo, H. Marchand, A. Rollet, in: Automatic test generation for data-flow reactive systems with time constraints, IFIP International Conference on Testing Software and Systems (ICTSS'10), Natal, Brazil, November, 2010.

[142] A. Hessel, K.G. Larsen, M. Mikucionis, B. Nielsen, P. Pettersson, A. Skou, in: Testing real-time systems using UPPAAL, Formal Methods and Testing (FORTEST), Springer, 2008, pp. 77–117.

[143] C. Gaston, R. Hierons, P. Le Gall, in: An implementation relation and test framework for timed distributed systems, IFIP International Conference on Testing Software and Systems (ICTSS'13), Istanbul, Turkey, November, 2013.

[144] A. En-Nouaary, M. Abu Talib, in: TTRTS: a tool for testing real time systems, International Conference on Sciences of Electronics, Technologies of Information and Telecommunications (SETIT'05), Tunisia, March, 2005.

[145] J. De Lara, H. Vangheluwe, in: AToM3: a tool for multi-formalism modeling and meta-modeling, Proceedings of the Fundamental Approaches to Software Engineering (FASE'02), LNCS 2306, 2002, pp. 174–188.

[146] N. Bertrand, A. Stainer, T. Jéron, M. Krichen, in: A game approach to determinize timed automata, Foundations of Software Science and Computation Structures (FOSSACS'11). Extended Version as INRIA Report 7381, Saarbrücken, Germany, March–April, 2011 http://hal.inria.fr/inria-00524830.

[147] K. El-Fakih, N. Yevtushenko, H. Fouchal, Testing timed finite state machines with guaranteed fault coverage, in: in: Workshop on Formal Approaches to Testing of Software (FATES'09), Eindhoven, The Netherlands, November, 2009.

[148] www.uppaal.com.

[149] H. Fouchal, E. Petitjean, S. Salva, in: An user-oriented testing of real time systems, IEEE/IEE Workshop on Real-Time Embedded Systems (RTES'01), December, IEEE Computer Society Press, London, UK, 2001.

[150] M. Mikucionis, K.G. Larsen, B. Nielsen, in: T-UPPAAL: online model-based testing of real-time systems, IEEE International Conference on Automated Software Engineering (ASE'04), September, IEEE Computer Society, Linz, Austria, 2004, pp. 396–397.

[151] K. G. Larsen, M. Mikučionis, and B. Nielsen, "Online testing of real-time systems using UPPAAL," in J. Grabowski and B. Nielsen, B. (Eds.) Workshop on Formal Approaches to Testing of Software (FATES'04), LNCS, 3395, Springer (2005) 79–94.

[152] A. Hessel, P. Pettersson, in: H.-D. Ehrich, K.-D. Schewe (Eds.), A test generation algorithm for real-time systems, Proceedings of the International Conference on Quality Software (QSIC), IEEE Computer Society Press, 2004, pp. 268–273.

[153] A. Hessel, P. Pettersson, in: L. Brim, B.R. Haverkort, M. Leucker, J. van de Pol (Eds.), Model-based testing of a WAP gateway: an industrial study, International Workshop on Formal Methods for Industrial Critical Systems and International Workshop on Parallel and Distributed Methods in Verification (FMIC/PDMC'06), LNCS 4346, Springer, 2007, pp. 116–131.

[154] M. Lahami, M. Krichen, M. Jmaiel, in: Selective test generation approach for runtime testing of dynamic behavioral adaptations, IFIP International Conference on Testing Software and Systems (ICTSS'15), Sharjah, Dubai, November, 2015.

[155] F. Cassez, A. David, E. Fleury, K.G. Larsen, D. Lime, in: Efficient on-the-fly algorithms for the analysis of timed games, Proceedings of the International Conference on Concurrency Theory (CONCUR'05), San Francisco, CA, August, 2005.

[156] G. Behrmann, A. Cougnard, A. David, E. Fleury, K.G. Larsen, D. Lime, in: UPPAAL-TIGA: time for playing games!, Proceedings of the International Conference on Computer Aided Verification (CAV'07), LNCS 4590, 2007, pp. 121–125.

[157] A. David, K.G. Larsen, S. Li, B. Nielsen, in: Timed testing under partial observability, IEEE International Conference on Software Testing, Verification, and Validation (ICST)" IEEE Computer Society. Denver, CO, April, 2009, pp. 61–70.

[158] A. David, K.G. Larsen, M. Mikučionis, O.L. Nguema Timo, A. Rollet, in: Remote testing of timed specifications, IEEE International Conference on Software Testing, Verification and Validation (ICST), LNCS 8254, 2013, pp. 65–81.

[159] B. Berthomieu, F. Vernadat, in: State class constructions for branching analysis of time Petri nets, Tools and Algorithms for the Construction and Analysis of Systems (TACAS'03), LNCS 2619, Springer, 2003, pp. 442–457.

[160] B. Berthomieu, F. Peres, F. Vernadat, in: K.S. Namjoshi, T. Yoneda, T. Higashino, Y. Okamura (Eds.), Model checking bounded prioritized time Petri nets, International Symposium on Automated Technology for Verification and Analysis (ATVA'07), LNCS 4762, Springer, 2007, pp. 523–532.

[161] B. Berthomieu, P.O. Ribet, F. Vernadat, The tool TINA—construction of abstract state spaces for Petri nets and time Petri nets, Int. JPR 42 (14) (2004).

[162] G. Frehse, K.G. Larsen, M. Mikučionis, B. Nielsen, in: Monitoring dynamical signals while testing timed aspects of a system, IFIP International Conference on Testing Software and Systems (ICTSS'11), Paris, France, November, 2011.

[163] G. Frehse, in: M. Morari, L. Thiele (Eds.), Phaver: algorithmic verification of hybrid systems past hytech, Hybrid Systems: Computation and Control (HSCC'05), LNCS 3414, Springer, 2005, pp. 258–273.

[164] H. Fouchal, E. Petitjean, S. Salva, in: An user-oriented testing of real time systems, IEEE/IEE Workshop on Real-Time Embedded Systems (RTES'01), IEEE Computer Society Press, London, UK, December, 2001.

[165] A. Khoumsi, A. En-Nouaary, R. Dssouli, M. Akalay, in: A new method for testing real time systems, Proceedings of the International Conference on Real-Time Computing Systems (RTCSA), Cheju Island, South Korea, December, 2000.

[166] A. Khoumsi, in: A method for testing the conformance of real time systems, IEEE International Symposium on Formal Techniques in Real-Time and Fault-Tolerant Systems (FTRTFT), Oldenburg, Germany, September, 2002.

[167] C. Jard, T. Jéron, in: TGV: theory, principles and algorithms, Proceedings of the 6th World Conference on Integrated Design & Process Technology (IDPT'02), Pasadena, CA, June, 2002.

[168] V. Rusu, in: Verification using test generation techniques, Formal Methods Europe (FME), LNCS 2391, 2002, pp. 252–271.

[169] D. Clarke, T. Jéron, V. Rusu, E. Zinovieva, in: STG: a symbolic test generation tool, Tools and Algorithms for the Construction and Analysis of Systems (TACAS), LNCS 2280, 2002, pp. 470–475.

[170] E.M. Clarke, O. Grumberg, D.A. Peled, Model Checking, MIT Press, Cambridge, MA, 1999.

[171] N. Een, N. Sorensson, in: An extensible SAT-solver, International Conference on Theory and Applications of Satisfiability (SAT'03), LNCS 2919, 2003, pp. 502–518.

[172] A. Macedo, W. Andrade, D. Rodrigues, P. Machado, in: Automating test case execution for real-time embedded systems, International Conference on Testing Software and Systems (ICTSS'10), Natal, Brazil, November, 2010.

[173] The FreeRTOS.org Project: FreeRTOS. http://www.freertos.org.

[174] J. Grossmann, D.A. Serbanescu, I. Schieferdecker, in: Testing embedded real time systems with TTCN-3, IEEE International Conference on Software Testing, Verification and Validation (ICST), Denver, CO, 2009.

[175] A. Khoumsi, in: Testing distributed real time systems using a distributed test architecture, Proceedings of the Sixth IEEE Symposium on Computers and Communications (ISCC), Hammamet, Tunisia, July, 2001.

[176] A. Khoumsi, in: Test execution for distributed real time systems, IEEE International Workshop on Communication Software Engineering (IWCSE), Marrakech, Morocco, December, 2002.

[177] A. Khoumsi, Testing distributed real-time systems in the presence of inaccurate clock synchronization, J. Inf. Softw. Technol. 45 (12) (2003) 853–864.

[178] A. Khoumsi, in: A centralized method for testing distributed real-time test systems, 10th IEEE North Atlantic Test Workshop (NATW), Gloucester, MA, May, 2001.

[179] A. Khoumsi, in: New results on testing distributed real-time reactive systems, IASTED International Conference on Parallel and Distributed Computing and Networking (PDCN), Innsbruck, Austria, February, 2002.

[180] A. Khoumsi, in: Testing distributed real-time systems: an efficient method which ensures controllability and optimizes observability, International Conference on Real-Time Computing Systems and Applications (RTCSA), Tokyo, Japan, March, 2002.

[181] A. Khoumsi, A temporal approach for testing distributed systems, IEEE Trans. Softw. Eng. 28 (11) (2002) 1085–1103.

[182] R.M. Hierons, M.G. Merayo, M. Nuñez, in: Using time to add order to distributed testing, International Symposium on Formal Methods (FM'12), LNCS 7436, Springer, 2012, pp. 232–246.

[183] R.M. Hierons, M.G. Merayo, M. Núñez, Timed implementation relations for the distributed test architecture, Distrib. Comput. 27 (3) (2014) 181–201.

[184] V. Garousi, L. Briand, Y. Labiche, in: Traffic-aware stress testing of distributed systems based on UML models, Proceedings of the International Conference on Software Engineering (ICSE'06), 2006, pp. 391–400.

[185] V. Garousi, L. Briand, Y. Labiche, Traffic-aware stress testing of distributed systems based on UML models, J. Syst. Softw. 81 (2) (2008) 161–185.

[186] V. Garousi, in: Traffic-aware stress testing of distributed real-time systems based on UML models in the presence of time uncertainty, IEEE Internatioanl Conference on Software Testing, Verification and Validation (ICST), Lillehammer, Norway, April, 2008.

[187] A. Avritzer, E.J. Weyuker, The automatic generation of load test suites and the assessment of the resulting software, IEEE Trans. Softw. Eng. 21 (9) (1995) 705–716.

[188] C.S.D. Yang, in: Identifying potentially load sensitive code regions for stress testing, Proceedings of the Mid-Atlantic Student Workshop on Programming Languages and Systems (MASPLAS), New Paltz, NY, April, 1996.

[189] J. Zhang, S.C. Cheung, Automated test case generation for the stress testing of multimedia systems, J. Softw. Pract. Exp. 32 (15) (2002) 1411–1435.

[190] L.C. Briand, Y. Labiche, M. Shousha, Using genetic algorithms for early schedulability analysis and stress testing in real-time systems, J. Genet. Program. Evolvable Mach. 7 (2) (2006) 145–170.

[191] M. Kwiatkowska, G. Norman, R. Segala, J. Sproston, Automatic verification of real-time systems with discrete probability distributions, Theor. Comput. Sci. 282 (1) (2002) 101–150.

[192] M.U. Uyar, S.S. Batth, Y. Wang, M.A. Fecko, Algorithms for modeling a class of single timing faults in communication protocols, IEEE Trans. Comput. 57 (2) (2008) 274–288.

[193] R.M. Hierons, M.G. Merayo, M. Nuñez, Testing from a stochastic timed system with a fault model, J. Log. Algebr. Program. 78 (2) (2009) 98–115.

[194] N. Wolovick, P.R. D'Argenio, H. Qu, in: Optimizing probabilities of real-time test case execution, IEEE International Conference on Software Testing, Verification and Validation (ICST'09), Denver, CO, April, 2009.

[195] E. Bayse, A. Cavalli, M. Núñez, F. Zaïdi, A passive testing approach based on invariants: application to the WAP, Comput. Netw. 48 (2) (2005) 247–266.

[196] C. Andrés, M.G. Merayo, M. Núñez, in: Passive testing of timed system, International Symposium on Automated Technology for Verification and Analysis (ATVA'08), LNCS 5311, Springer, 2008, pp. 418–427.

[197] M.G. Merayo, M. Núñez, in: Passive testing of stochastic timed systems, IEEE International Conference on Software Testing, Verification and Validation (ICST'09), Denver, CO, 2009, pp. 71–80.

[198] M.W. Aziz, S.A.B. Shah, in: Test-data generation for testing parallel real-time systems, IFIP International Conference on Testing Software and Systems (ICTSS'15), Sharjah, Dubai, November, 2015.

[199] N. Binkert, B. Beckmann, G. Black, S.K. Reinhardt, A. Saidi, A. Basu, J. Hestness, D.R. Hower, T. Krishna, S. Sardashti, et al., The gem5 simulator, ACM SIGARCH Comput. Archit. News 39 (2) (2011) 1–7.

[200] A. Hessel, K.G. Larsen, B. Nielsen, P. Pettersson, A. Skou, in: Time-optimal real time test case generation using UPPAAL, Proceedings of the Formal Approaches to TEsting of Software (FATES'03), LNCS 2931, Springer, 2003, pp. 114–130.

[201] G. Behrmann, A. Fehnker, T. Hune, K.G. Larsen, P. Pettersson, J. Romijn, in: T. Margaria, W. Yi (Eds.), Efficient guiding towards cost-optimality in UPPAAL, Tools and Algorithms for the Construction and Analysis of Systems (TACAS'01), LNCS 2031, Springer-Verlag, 2001, pp. 174–188.

[202] A. Arcuri, M.Z. Iqbal, L. Briand, in: Black-box system testing of real-time embedded systems using random and search-based testing, IFIP International Conference on Testing Software and Systems (ICTSS'10), Natal, Brazil, November, 2010.

[203] T.Y. Chen, F. Kuoa, R.G. Merkela, T. Tseb, Adaptive random testing: the art of test case diversity, J. Syst. Softw. 83 (1) (2010) 60–66.

[204] J. Peleska, A. Honisch, F. Lapschies, H. Loeding, H. Schmid, P. Smuda, E. Vorobev, C. Zahlten, in: A real-world benchmark model for testing concurrent real-time systems in the automotive domain, IFIP International Conference on Testing Software and Systems (ICTSS), Paris, France, November, 2011.

ABOUT THE AUTHORS

Rachida Dssouli is professor and founding Director of Concordia Institute for Information Systems Engineering (CIISE), Concordia University. She received the Doctorat d'Université degree in Computer Science from the Université Paul-Sabatier of Toulouse, France, in 1981, and the Ph.D. degree in Computer Science in 1987, from the University of Montréal, Canada. She has been a professor at the Université Mohamed 1er, Oujda, Morocco, from 1981 to 1989, Assistant professor at the Université de Sherbrooke, from 1989 to 1991, and Full professor at the Université de Montréal until May 2001. Her research area is in Testing and Verification, Communication software engineering, Requirements engineering and Service Computing. Ongoing projects include Model based Specification and Testing of Avionics Software, Service Composition, Conformance Testing based on FSM/EFSM and Timed Automata, Live Testing, and Verification of Real-Time Systems.

Ahmed Khoumsi received the Engineer degree in Aeronautics and Automation from the engineer school SUPAERO (Toulouse, France) in 1984. In 1988, he received the Ph.D. degree in Robotics and Automation from the University Paul Sabatier in Toulouse. From 1984 to 1988, he conducted his research activities in the LAAS, a CNRS research center, in Toulouse. From 1989 to 1992, he was an Assistant Professor in Robotics and Computer Engineering at the engineer school ENSEM (Casablanca, Morocco). From 1993 to 1996, he was a Postdoctoral Fellow in the Communication Protocols group of the University of Montreal. From 1996 to June 2000 he was Assistant Professor, from June 2000 to June 2006 he was Associate Professor, and since June 2006 he is Full Professor, all in the Department of Electrical and Computer Engineering, at the University of Sherbrooke, Sherbrooke, QC, Canada. He was a 1-year Visiting professor at the research institute IRISA (Rennes, France) and at the King Saud University (Riyadh, Saudi Arabia), in 2002–2003 and 2010–2011, respectively. His present research activities include: supervisory control, diagnosis, and prognosis of discrete event systems; design and creation of web services; design and analysis of firewall security policies; and intelligent transportation systems.

Mounia Elqortobi is a Ph.D. degree student in Information Systems Engineering, Concordia Institute for Information Systems Engineering. She received her Master of Engineering degree in Quality Systems Engineering from Concordia University, in 2015. She received her bachelor degree in Computer Science from Computer Science and Software Engineering Department, Concordia University, in 2010. She has more than 6 years of work experience in software and systems quality assurance. Her interests include: model-based testing, model checking, software engineering, quality assurance in software and systems engineering, and testing and verification.

Jamal Bentahar received his Ph.D. degree in computer science and software engineering from Laval University, Canada, in 2005 and his Master in Software Engineering from École Nationale Supérieure d'Informatique et d'Analyse des Systèmes, Morocco in 1999. He is a Full Professor with Concordia Institute for Information Systems Engineering, Faculty of Engineering and Computer Science, Concordia University, Canada. From 2005 to 2006, he was a postdoctoral fellow with Laval University, and then an NSERC PDF fellow at Simon Fraser University, Canada. His research interests include services computing, applied game theory, computational logics, model checking, multiagent systems, and software engineering.

CHAPTER FOUR

Advances in Testing Software Product Lines

Hartmut Lackner*, Bernd-Holger Schlingloff†
*Model Engineering Solutions GmbH, Berlin, Germany
†Humboldt–Universität and Fraunhofer FOKUS, Berlin, Germany

Contents

1.	Introduction	158
2.	Specification of Variability	160
	2.1 Feature Modeling	162
	2.2 Variability Modeling	164
	2.3 A Basic Variability Language	165
3.	Model-Based Testing for Product Lines	170
	3.1 Application-Centered and Product-Centered Test Generation	170
	3.2 An Evaluation of Product Line Test Generation	178
4.	Assessment of Product Line Test Suites	183
	4.1 Potential Errors in Model-Based Product Line Engineering	184
	4.2 Mutating Domain Models	187
	4.3 Evaluation	191
5.	Test-Driven Product Sampling	193
	5.1 Sampling Configurations From Generic Test Cases	194
	5.2 Evaluation	198
6.	Assignment of Product Line Test Cases	200
	6.1 Colored Test Cases	201
	6.2 Automated Test Coloring via Model Checking	204
7.	Related Work	206
	7.1 Specification of Software Product Lines	206
	7.2 Model-Based Test Generation	206
	7.3 Feature Modeling	207
	7.4 Mutation Analysis	207
	7.5 Sampling of products	208
	7.6 Test case assignment	208
8.	Future Developments	209
9.	Summary and Conclusion	210
	References	211
	About the Authors	216

Advances in Computers, Volume 107
ISSN 0065-2458
http://dx.doi.org/10.1016/bs.adcom.2017.07.001

157

Abstract

In this chapter, we describe some recent techniques and results in model-based testing of software product lines. Presently, more and more software-based products and services are available in many different variants to choose from. However, this brings about challenges for the software quality assurance processes. Since only few of all possible variants can be tested at the developer's site, several questions arise. How shall the variability be described in order to make sure that all features are being tested? Is it better to test selected variants on a concrete level, or shall the whole software product line be tested abstractly? What is the quality of a test suite for a product line, anyway? If it is impossible to test all possible variants, which products should be selected for testing? Given a certain product, which test cases are appropriate for it, and given a test case, which products can be tested with it?

We address these questions from an empirical software engineering point of view. We sketch modeling formalisms for software product lines. Then, we compare domain-centered and application-centered approaches to software product line testing. We define mutation operators for assessing software product line test suites. Subsequently, we analyze methods for selecting product variants on the basis of a given test suite. Finally, we show how model checking can be used to determine whether a certain test case is applicable for a certain product variant.

For all our methods we describe supporting tools and algorithms. Currently, we are integrating these in an integrated tool suite supporting several aspects of model-based testing for software product lines.

1. INTRODUCTION

Due to increasing market diversification and customer demand, most industrial products today are available in many different variants. For example, when buying a new car, one can choose not only between different brands and makes. Within a model, one has options on the type and power of the engine, on the navigation and entertainment system, on various driver assistance functions, etc. There is a common basis on which these different variants are built (chassis, doors, power train, and other elements).

A *product line* (or *product family*) is a set of products offered by a producer to customers, which have the same base functionality and share a common set of base elements. The members of a product line differ in the *features* which they offer to the customer. Thus, the individual products in a product line have a similar "look–and–feel"; however, they differ in that one product may offer more or other functionality than another one.

In software-based systems, these features are realized via software. The concept of a software product line originates in the work of Parnas [1].

It has gained much attention by the research and consultancy of the Carnegie Mellon University Software Engineering Institute [2]. The CMU SEI defines a *software product line* to be a "set of software-intensive systems that share a common, managed set of features satisfying the specific needs of a particular market segment or mission and that are developed from a common set of core assets in a prescribed way" [3].

Software product lines are abundant in today's cyber physical systems; most electronic control units, e.g., in cars or trains, are configurable or come in multiple variants, as well as software in consumer products like coffee machines, dishwashers, mobile phones, etc. Also software products itself like, e.g., the ordering system of a web shop, are available in different variants to fit varying customer needs.

A challenge common to the development of these systems is that the software is similar, but not identical for all products; there are slight differences according to the features exhibited by a particular product. Sources of variability include planned diversity for different user groups, evolution and enhancement of products, and reuse of modules from one product in another one. Software product line engineering addresses this challenge. The main goal of software product line development is the strategic reuse of software artifacts. There have been various approaches to reuse: by copy and paste, macros, subroutines, modules, objects, components, and services. The common problem in all of these approaches is that reuse increases the probability of errors. Therefore, quality assurance for software product lines is of utmost importance.

In this chapter, we summarize and extend some of our previous work on this topic. We address several questions related to the model-based testing of software product lines. First, we address the problem of how to *specify* a software product line. We describe feature models and other artifacts that arise in a model-based development of software product lines.

We then consider the question of how to *generate* test cases from models. Basically, there are two competing approaches: one can derive individual test suites for individual products from an appropriate instantiation of a domain model. Or, one can try to generate a generic test suite from the domain model and instantiate this generic suite for each individual product. We formalize and compare these two approaches and derive guidelines on when to use which one.

Given a test suite, an important issue is to asses its *error detection capability*: how likely will the test suite uncover an error, given a faulty system under test? We investigate this question for the automatically generated test suites from the previous section. We define mutation operators, an experiment setup, and a tool suite to determine the mutation score for a test suite.

For a large product line, there may be zillions of different variants; it is impossible to test all of them. The question is, which products should be selected for testing the product line? We suggest that the products should be sampled using the test suite generated from the domain model. Moreover, we suggest that in this sampling, certain optimization criteria should be taken into consideration, such as "the selected products should be as diverse as possible." We sketch an algorithm based on Boolean optimization for this sampling and evaluate the effectiveness of the approach with fault injection techniques.

In many companies, especially when transitioning to a product line development process, there exists a large body of test cases. In order to be able to reuse these test cases for new products, we sketch a model–checking algorithm, which determines whether a certain test case is applicable for a certain product. We use a three–valued coloring scheme to distinguish between positive tests, negative tests, and tests which are insignificant for a certain product. We describe a prototypical implementation of our coloring algorithm and its application.

Throughout the chapter, we use the example of a web shop system to illustrate our ideas.

This chapter is structured as follows: After the introduction in Section 1, we define modeling formalisms for software product lines in Section 2. Subsequently, we show how to use domain models for test case generation in Section 3. Then, in Section 4, we describe a mutation system for assessing the error detection capability of a product line test suite. Section 5 deals with the problem of how to sample products for testing, and Section 6 with the problem of how to reuse test cases for different products. Finally, in Section 7 we discuss related literature, and, in Section 8, some outlook on future work.

2. SPECIFICATION OF VARIABILITY

Two main driving factors for software product line engineering are the growing customer expectations for individualized products, and the potential to reuse existing software assets in the design of a product. The objective is to increase the number of product features while keeping the overall system engineering costs at a reasonable level. Product line development usually involves two engineering processes: *domain engineering* and *application engineering*. In domain engineering, reusable components are developed by a domain analysis and a domain reference architecture. In this process, commonalities and variabilities are analyzed and generic artifacts are developed. In application engineering, actual products serving different customer and market needs are derived from the domain artifacts. Products are built by instantiating and composing generic artifacts from the domain engineering process. Thus, domain artifacts are reused to exploit the product line's commonalities. The instantiation and composition should be largely automatic. This way, many different products can be derived in an efficient way. There should be also feedback loops from application engineering to domain engineering such that updates from different individual products can be generalized and adapted to the product line.

Model-based design is a particular form of software development, where a system model is continuously used as the central artifact throughout the whole engineering process. Initially, requirements are captured in an abstract model, e.g., in SysML, representing the system specification. This abstract model is refined and transformed into a concrete implementation model, e.g., in UML or Simulink$^{®}$. From this implementation model, executable code is generated automatically by a suitable model compiler.

For model-based software product line engineering, the artifacts produced during domain engineering are mostly models. However, these models are generic, allowing an instantiation into different product models. During application engineering, the product models are refined and compiled as in "ordinary" model-based design.

Fig. 1 depicts the model-based domain and application engineering work flows and their interrelations. The domain engineering process is generic,

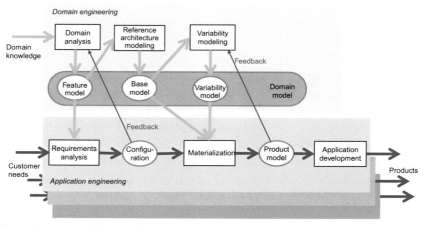

Fig. 1 Various models in software product line engineering.

whereas the application engineering process exists in several instances, namely one for each product. Tasks are denoted by rectangles, and the models which are the results of tasks are given in circles.

2.1 Feature Modeling

There have been various attempts to specify the variability of products belonging to a product line. Most approaches are based on a so-called feature model. Feature models facilitate the explicit design of global system variation points [4].

A *feature* is the description of a designated functionality which is visible to the customer and forms an added value to the product. Each feature has a unique name and represents one characteristics of a product which is interesting for the stakeholder, e.g., a special added value. A *feature model* is a structuring concept for a set of features; it forms an explicit description of commonalities and differences of various products. Feature models are usually organized as and-or-trees, where each node is marked with the name of a feature. The root of the feature tree is the name of the product family. A parent feature can have the following relations to its subfeatures:

- *Mandatory*: child feature is required,
- *Optional*: child feature is optional,
- *Or*: at least one of the children features must be selected, and
- *Alternative*: exactly one of the children features must be selected.

Additionally, it is allowed to attach Boolean constraints on features to the feature tree. For example, one may specify additional (cross-tree) constraints between two features A and B: (i) A *requires* B: the selection of A implies the selection of B, and (ii) A *excludes* B: both features A and B must not be selected for the same product. Tools for maintaining feature trees include pure::variants®, BigLever Gears®, FeatureIDE [5], and EASy-Producer [6].

Fig. 2 shows the feature model of our web shop example which is used throughout this chapter. The eShop is a fictional e-commerce web shop designed for this chapter; it covers only a small part of the "real" web application. It provides the following functionality: each eShop contains a catalog in which the available goods are listed. A customer can browse the catalog of items and put orders in the cart. Once the customer is finished, he/she can checkout and may choose from up to three different payment options, depending on the eShop's configuration. Payment can be via bank transfer, credit card, or eCoins. Some variants may provide a search function within the catalog. The transactions are secured by either a standard or high security

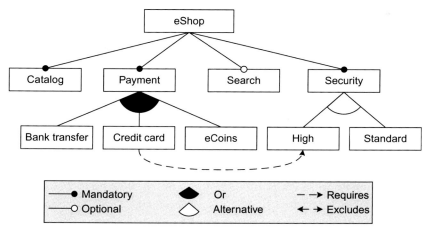

Fig. 2 A feature model for the eShop example.

server. Credit card payment is only offered if the eShop also implements a high security server.

Feature models can also be represented in propositional logic: for each feature model, there is an equivalent Boolean formula. The propositions in this formula are the feature names, and the logical connectives are derived from the tree structure (see, e.g., Refs. [7, 8]). For the eShop feature model in Fig. 2, a Boolean formula representing the feature tree is

$FM = (eShop$
$\quad \wedge (eShop \rightarrow Catalog) \wedge (eShop \rightarrow Payment) \wedge (eShop \rightarrow Security)$
$\quad \wedge (Payment \rightarrow BankTransfer \vee CreditCard \vee eCoins)$
$\quad \wedge (Security \rightarrow (High \wedge \neg Standard) \vee (\neg High \wedge Standard))$
$\quad \wedge (Catalog \rightarrow eShop) \wedge (Payment \rightarrow eShop)$
$\quad \wedge (Search \rightarrow eShop) \wedge (Security \rightarrow eShop)$
$\quad \wedge (BankTransfer \rightarrow Payment) \wedge (CreditCard \rightarrow Payment)$
$\quad \wedge (eCoins \rightarrow Payment) \wedge (High \rightarrow Security) \wedge (Standard \rightarrow Security)$
$\quad \wedge (CreditCard \rightarrow High))$

Since our feature model involves optional features, there are several potential products. Only a few of these will be materialized as actual products in the market. Given any feature model, a *resolution model* (or simply *resolution*) is an assignment of truth values to feature names. The assignment of a value to a proposition indicates whether the corresponding feature is selected (*true*) or deselected (*false*). A resolution is called *valid*, if the corresponding Boolean formula evaluates to true. A valid resolution is also called a *product configuration*.

For instance, the following two assignments *P1* and *P2* are product configurations for the eShop feature model presented in Fig. 2.

$P1 \ = eShop, Catalog, Payment, BankTransfer, \neg CreditCard, \neg eCoins,$
$\quad\quad \neg Search, Security, \neg High, Standard$
$P2 \ = eShop, Catalog, Payment, BankTransfer, CreditCard, eCoins,$
$\quad\quad Search, Security, High, \neg Standard$

It can be seen that both configurations satisfy the formula *FM* above. *P2* is a *maximal* configuration with respect to the number of features. For any feature model with n features, there are between 0 and 2^n product configurations. In particular, the eShop feature model allows 20 different configurations.

2.2 Variability Modeling

Although a feature model captures the system's variation points in a concise form, its elements are only propositional symbols [9]. Their semantics has to be provided by mapping them to other development artifacts. Such a mapping can be defined either explicitly, in a separate variability model, or implicitly, within the referenced artifacts. In the following, we introduce the three major paradigms for variability modeling:

2.2.1 Annotative Variability Modeling
A base model contains every element that is used in at least one product configuration and, thus, subsumes every possible product [11] (Fig. 3A). Subsequently, model elements are removed to resolve a valid variant.

2.2.2 Compositional Variability Modeling
In contrast to annotative languages, compositional languages start from a minimal core that contains features that are common to all possible products. From this starting point additional features will be added by a designer (Fig. 3B).

2.2.3 Transformational Variability Modeling
In transformational variability, compositional and annotative methods are combined. Model elements can be removed and added to resolve a variant. A well-known approach for this is so-called delta modeling (also delta-oriented programming) [12]. Delta modeling consists of two parts: the first

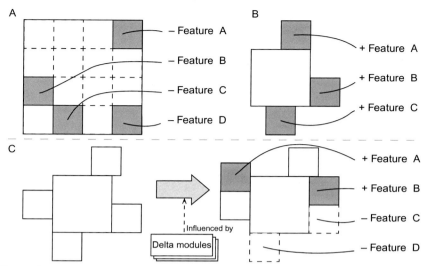

Fig. 3 Annotative (A), compositional (B) variability (based on Ref. [10]), and delta modeling (C).

one is the design of a core product comprised of a set of feature selections that represent a valid product. The second part is the specification of a set of so-called delta modules which describe changes (deltas) to the core module. Each delta is either the construction (add) or destruction (remove) of an element from the product model (Fig. 3C). A delta module then is associated to one or more features. Whenever a feature is selected or deselected the associated deltas are applied to the product model.

2.3 A Basic Variability Language

In this chapter, we describe a custom modeling language for annotative variability. This language is inspired by the "Common Variability Language" (CVL) standardization initiative [13].

As a concrete syntax for feature models, we provide a meta-model defined in Ecore with the Eclipse modeling framework (EMF) as depicted in Fig. 4. This meta-model offers basic functionalities for designing mandatory, optional, or, alternative, and binary cross-tree constraints between features.

We use an explicit *variability model* for mapping features to elements of a *base model*. That is, a base model is the artifact which implements the features of the product line. (The base model sometimes called "the 150% model," a

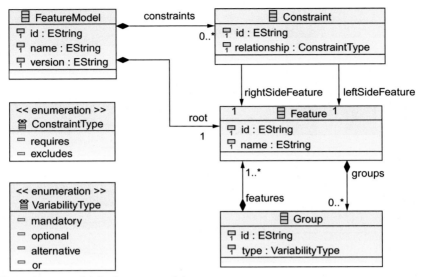

Fig. 4 Meta-model for feature models.

terminology which we refrain to adopt.) Each feature corresponds to some parts of the base model.

Formally, a base model can be any model which is an instance of some MOF meta-model. In our work, a base model is a UML model consisting of state machine diagrams, class diagrams, and OCL formulae. The base model describes realizations for *all* features; thus, if the feature model contains conflicting features, then the base model may not represent a possible product.

Part of the base model of the eShop example, the UML state machine diagram, is given in Fig. 5. This state machine is designed from a user's perspective. Whenever a user starts the eShop, the system initializes and waits for the user in state `Start` to open the product catalog. Upon receipt of the signal `openProductCatalog`, it invokes the submachine `ProductSelection`, which initially outputs a list of product items and then waits for further user input in state `Catalog`. The system sends outputs to the user or known banks by instantiating the output message, e.g., `listProducts o;` and sending it via calling the method `send` of its outport object. The eShop example has exactly one outport, called `Out`; hence, a message object `o` is sent by calling `Out.send(o)`. The user can display the product details by sending the message `productDetails` to the system. Not shown is the necessary parameter id of the product for which the details shall be displayed. When the system is

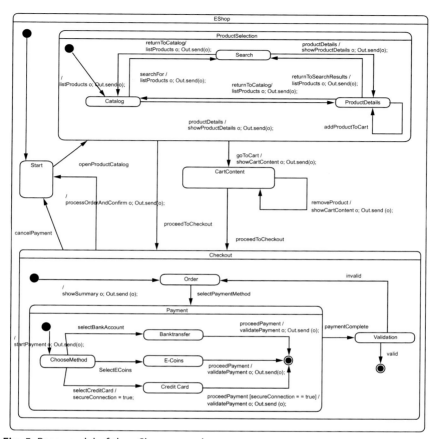

Fig. 5 Base model of the eShop example.

in state ProductDetails, the user may add the displayed product to his cart and/or return to the catalog. If enabled, the user can search for a particular product in state Catalog by sending signal searchFor, which will return a list of products that match the search term entered by the user. Again, the user may view the product details and add the product. It is also possible to return to the search or the catalog.

Anytime state ProductSelection is active, the user may view the cart's contents (goToCart) or proceed to checkout (proceedToCheckout). While the user displays the contents of the cart, items can be removed (removeProducts) before proceeding to the checkout (proceedToCheckout). When the user proceeds to the checkout the eShop invokes the submachine Checkout. Here, the user is presented with a final summary before selecting a payment method. Depending on the configuration of the eShop, the user may select from up to three payment methods. The internal workings of the individual payment methods are not modeled. When a customer has entered his payment data, s/he may proceed to the validation (proceed-Validation). Credit card payments will not be validated unless the secureConnection flag is set to true. This flag is raised whenever a high security eShop product is built. The eShop will send the payment data to the corresponding bank for validation (validatePayment). The bank acknowledges the receiving of the data by sending a paymentComplete signal to the eShop. If the data provided by the customer is correct and the transaction is carried out, the bank sends the message valid, and otherwise invalid. If a valid signal is received by the eShop it will continue to process the order and lead the user back to the state Start. Otherwise, if the signal invalid is received, the user is prompted again to choose the payment method and data. The user can abort the checkout process at any time by sending a cancelPayment signal to the eShop.

The *variability model* maps features to elements of the base model. In Fig. 6, we give a meta-model for the variability modeling language as implemented with Ecore. The language refers to concepts of *elements* from the UML and to *features* from our feature modeling language. Multiple

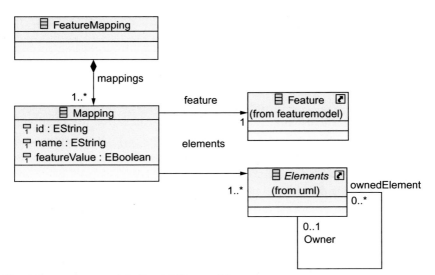

Fig. 6 Ecore meta-model of variability models.

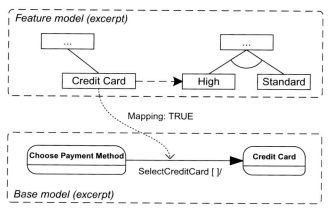

Fig. 7 Software product line design with annotative variability.

features mapping to the same base model element are interpreted as a conjunction of features. Additionally, each mapping has a Boolean flag that indicates whether the mapped model elements are part of the product when the feature is selected (*true*) or unselected (*false*).

Fig. 7 shows an excerpt of the eShop variability model, where parts of the feature model are depicted in the upper half and parts of the state machine's payment process are shown in the lower half. In between is the mapping, denoted by a dotted edge, from feature *Credit Card* to the transition labeled as `SelectCreditCard[]/`.

The complete mapping in the variability model is as follows:

- *Catalog*, *Payment*, and *Security* are mandatory and need no mapping.
- *Bank Transfer* maps to all transitions connected to state `Bank Transfer`.
- *eCoins* maps to all transitions connected to state `eCoins`.
- *Credit Card* maps to all transitions connected to state `Credit Card`.
- *High* maps to attribute secureConnection with its default value set to true.
- *Standard* maps to attribute secureConnection with its default value set to false.
- *Search* maps to all transitions connected to state `Search`.

We refer to the triple of feature model, base model, and variability model as the *domain model*. From such a domain model, product models can be resolved for a given product configuration. In a model-based design process, implementation code or test cases can be derived automatically or semiautomatically from these product models.

3. MODEL-BASED TESTING FOR PRODUCT LINES

Systematic testing is the most common method for quality assurance of complex software. A *system under test* (SUT) is executed with a systematically constructed set of test cases, the *test suite*, and the observed system behavior is compared with the expected one. The quality of a test suite is often measured in terms of requirements and/or code coverage. Constructing a test suite which satisfies given coverage criteria by hand may be a tedious exercise. The term *model-based testing* subsumes various techniques in which test suites are constructed from a given model [14].

There are several such techniques: in some approaches test case models, e.g., use case diagrams, are used for test case description and test implementation [15]. In other approaches, e.g., the classification tree method, test data is generated from test data models [16]. In this chapter, we consider the automation of test design, where a test suite is generated from a dedicated test model. The test model contains test–relevant information about the intended behavior of the system under test and/or the behavior of its environment. Each generated test case is a sequence of test stimuli and expected system reactions. A test case may also pose preconditions on the system's state or configuration that must be fulfilled before the test case can be executed. Additionally, it may specify postconditions which must hold after the execution of the test case.

For software product lines, there are various possibilities to define model-based testing processes. One approach is to resolve a representative set of products from a product line for the purpose of testing, and then to generate test cases for these products. Another possibility is to generate tests from domain–level artifacts, and to specialize this generic test suite to specific products. Subsequently, we elaborate both approaches, and compare them. This material is based on the forthcoming dissertation of the first author [17]. We have implemented and evaluated both approaches in a prototypical tool "SPLTestbench"; we report on the results using several examples.

3.1 Application-Centered and Product-Centered Test Generation

A test generation method for software product lines faces two challenges: covering a significant subset of products, and covering a significant subset of the test focus on the domain level. As the products in a product line share

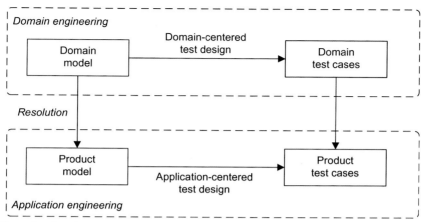

Fig. 8 Product line testing.

commonalities, there may be significant overlap in test suites, and test cases may be applicable to more than one product.

As described in the previous section, a domain model consists of a base model (e.g., a UML state machine), a feature model explicitly expressing the product line's variation points (e.g., a feature tree), and a variability model connecting these two. From domain artifacts, application artifacts (products and tests) can be derived. Based on this, we define two approaches to automated test design for software product lines. These are depicted in Fig. 8: *application-centered*(AC) and *domain-centered*(DC) testing.

In the AC approach, first a representative set of products for testing is selected. Then, test cases are generated from each of these models. The focus of the test generation process is on satisfying a defined coverage for each test model. This may lead to an overlap in the resulting test suites.

In contrast to this, in the DC approach the domain model is directly used for test design. In this approach, the focus is on the behavior defined at the domain level; coverage criteria of single products are not considered. There is still variability in the choice of the concrete products for which the test cases will be executed. Both approaches are investigated in more detail in the following paragraphs.

3.1.1 Application-Centered Test Design

Any test design method that binds the variability by selecting products before the test design phase can be called application–centric (see Fig. 9). Usually, it is not feasible to build and test *all* possible products of a product line.

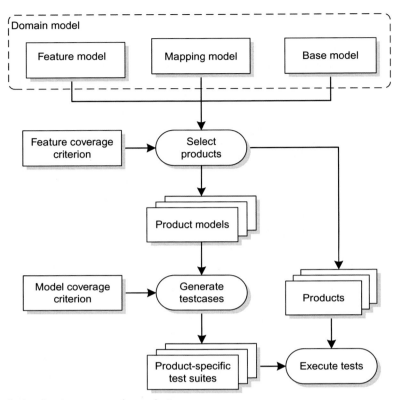

Fig. 9 Application-centered test design process.

Therefore, in application–centered test design, product models to be tested are selected from the domain model before the test design starts. The selection is done according to a predefined *feature coverage criterion*. For example, pairwise feature coverage (see Refs. [18, 19]) is achieved when for each tuple of features all valid combinations of feature presence and absence are represented by at least one product model. Tests are performed only on the products corresponding to the selected product models.

Using conventional test generation methods, test suites can be automatically obtained from the selected product models. The test generator can be configured such that is satisfies a certain *model coverage criterion*. In our example, this may be any coverage criterion applicable to UML state machines, e.g., state or transition coverage.

Since each test case is generated for one configuration, the resulting test suite is specific to its respective product. Therefore, the test generation yields *application-specific test suites.* Due to the commonalities of the products of a product line, a test case that was generated for one configuration may be applicable to other configurations as well. Consequently, test cases that aim for the same goals are executed over and over again. Therefore, there may be significant redundancies in application–centered test suites.

3.1.2 Domain-Centered Test Design

In order to avoid these potential drawbacks, in domain–centered test design domain engineering level artifacts are used for test generation. This approach preserves the variability until a product has been selected for test execution.

A major advantage of this approach is that one can focus on testing aspects of the domain model, without having to derive products first. Thus, the overlap of the results from independently generating tests for similar products can be avoided. The coverage of test targets can be maximized, which leads to high-quality test cases.

For domain–centered test design, all domain artifacts should be taken into consideration. Using a base model as the only source of information is not adequate. The base model lacks information about the features, and about the associations of model elements to features. However, this information is important for a test generator: features impose additional constraints on the behavior of the system. Since the base model contains implementation information for *all* features, it may include contradictory requirements. It might not even be a correct model according to the UML syntax. Thus, a challenge of this approach is to merge a domain model, consisting of a feature model, a variability model, and a base model, into a single model artifact that a standard test generator will accept as valid input (Fig. 10). Subsequently, we describe two solutions to this problem: (1) the *step-by-step* approach: sequentially excluding nonconforming configurations during test design time, and (2) the *preconfiguration* approach: choosing a valid configuration before the design of individual test cases.

- *The step-by-step approach:* The key idea of the step-by-step approach is to sustain the variability until it becomes necessary to bind it. Therefore, at the beginning of each test case design the test case is applicable to any valid product of the product line. Since not necessarily all valid paths in the base model are applicable to all products, the test designer must take account of test steps that bind variability. A test step must bind variability if not all products do conform. Subsequently, the set of valid products for this particular test case must be reduced by the set of

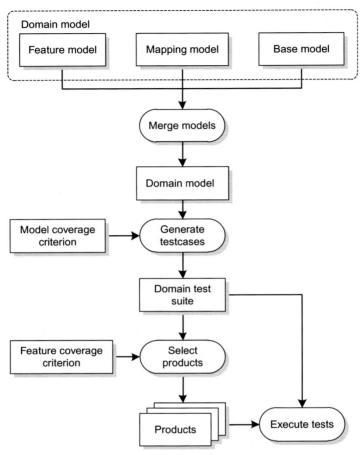

Fig. 10 Domain-centered test design process.

nonconforming products. Hence, each test case is valid for any of the remaining products that do conform.

We implemented the step-by-step approach for state machines as follows: The tracking of the excluded products can be achieved by introducing a Boolean variable into the system class for each feature that is not a core feature (*feature variable*). This variable is set whenever a transition added to a test case forces the mapped feature to be present (*true*) or prohibits its presence (*false*). For preventing repeated assigning to such a feature variable, an additional control variable is necessary. Therefore, another Boolean variable is added for each noncore feature to the system class (*control variable*) and must be initialized with *false*. Each of these variables tracks whether the corresponding feature has not yet been set and is thus free (*false*) or was already set (*true*). In the latter case, no further assignments to the feature variable are allowed as the feature is bound to the value of the corresponding feature variable.

The guards and effects on the transitions of the respective state machine can then be instrumented with these variables to include variability information in the state machine. For each feature f_i that is mapped by a mapping $m_{f_i,t}$ to a transition t, its partial feature formula $pf_{f_i,t}$ is derived: a partial feature formula for a particular feature f is constructed as a conjunction of (i) the feature f itself, (ii) f's *parents* p, and (iii) every feature r that is *required* by f. Depending on the structure of the feature model, not all features related to f are captured by this approach. Any parent or required feature may require additional features. Therefore, steps (ii) and (iii) are repeated for every p and r until the formula is stable. So far, the formula contains features that must be selected for f, but since some of the selected features may require the absence of other features, we combine the formula with another conjunction of (iv) features *excluded* by already selected features and (v) *alternative* features. Finally, the formula can be reduced by removing all core features, because they are a mandatory part of every product. For instance, the partial feature formula for the CreditCard feature from our eShop example is

$$F = CreditCard \wedge High \wedge \neg Standard$$

Since we have now derived all features that have to be accounted for before taking transition t, we collect them in a single conjunction:

$$G_t = \bigwedge_{i=1}^{n} pf_{f_i,t}$$

We still have to incorporate the protection against repeated writings by substituting each feature literal in G_t with the following expression: $(\neg f_c \vee (f_v == m_{f,v}))$, where f_c is the control variable of feature f, f_v is the feature variable of f, and $m_{f,t}$ is the value of the feature mapping's flag associated with transition t. The resulting expression can safely be conjoined with t's original guard.

Finally, t's effect must bind the variability of all associated features. This is possible by setting the control variable f_c to true and the feature variable f_v to the value of its mapping's flag for each feature that appeared

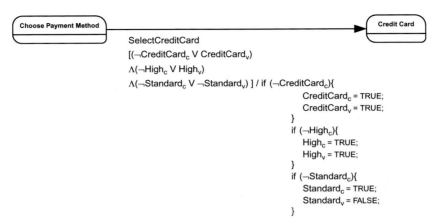

Fig. 11 Excerpt of the merged domain model with step-by-step approach applied.

in G_t. Thus, for each feature f in G_t we append the following code to the effect of transition t:

if $\neg f_c$ **then**
$$f_c \leftarrow \top$$
$$f_v \leftarrow m_{f,t}$$

Once the test generator executes this code, the feature is bound and it is not possible to change the binding for this test case anymore. Fig. 11 shows the result of merging the domain model into a single UML state machine for the excerpt of the eShop introduced in Fig. 7.

After test generation has finished, the valid configurations for a particular test case can be read from the feature variables in each test case. Since the test cases may contain variability we obtain an *incomplete configuration* from each test case. An incomplete configuration is a configuration that supports a three–valued semantics for features instead of two values. The first two values are the same as in normal configurations (selected/unselected), the third stands for *undecided*. An undecided feature expresses variability by making no premise on the presence of the feature. Hence, each of the resulting test cases is generic for any product of the product line that conforms to the following: for each control variable that is evaluated to true, the corresponding feature variable evaluation indicates whether this feature must be selected or unselected in the product. Features for which the respective control variable evaluates to *false* are yet undecided and thus not evaluated.

- *The Preconfiguration Approach:* In the preconfiguration approach, test goals are selected from the domain model and also the test design is performed on this model similar to the step-by-step approach. However, during the design of an individual test case, the product configuration is fixed from the beginning of each test case and must not change before a new test case is created. Consequently, within a test case the test designer is limited to test goals that are specific to the selected product configuration. Thus, satisfying all domain model test goals is a matter of finding the appropriate configurations.

  We implemented the preconfiguration approach by adding a configuration signal to the very beginning of the base model. To this end, we introduce a new state to the state machine, redirect all transitions leaving the initial state to leaving this new state, and add a transition between the initial state and the new state. Due to the UML specification the redirected transitions must not have a trigger, which is why we can add a trigger for configuration purposes to each of them. The trigger listens to a configuration signal that carries a valuation for all noncore features. The guard of these transitions must protect the state machine to be configured with invalid configurations and thus contains the propositional formula corresponding to the product line's feature model. Since any configuration that is provided by the signal must satisfy the guard's condition, only valid configurations are accepted.

  After validating the configuration, the parameter values of the signal will be assigned to system class variables by the transition's effect. Hence, for each noncore feature a Boolean variable, indicating whether the feature is selected or not, is added to the system class. Again, transitions specific to a set of products are protected by these variables. This is similar to the step-by-step approach, where the base model behavior is limited to a potential behavior of an actual product. However, control variables need not to be checked during test design, since the configuration is fixed and valid from the beginning of each test case. Therefore, it is sufficient to derive the partial feature formulas *pf* for all features f_n that are mapped to a transition t by a mapping $m_{f_i,t}$ and construct a conjunction from these formulas:

$$G_t = \bigwedge_{i=1}^{n} pf_{f_i,t}.$$

Fig. 12 Excerpt of the merged domain model in the preconfiguration approach.

For conjoining G_t with t's guard, the feature literals must be exchanged by the corresponding feature variables from the class. Fig. 12 depicts the resulting merged domain model for this approach. As a result, no product can conform to any test case's first step, since it was used to set the configuration and does not present the real system's behavior. In a simple postprocessing action, this configuration step can be removed from the test cases before testing is performed.

With these transformations to the base model, a test designer can create test cases for the product line. However, each test case will be specific to one configuration. In order to create generic test cases which are applicable to more than one product, we can apply a model transformation.

The additional transformation steps consist of adding Boolean control variables for each noncore feature to the system class. These control variables are initialized with *false*. Effects on transitions are added which change them to *true* when traversed by the test generator. More precisely, for every transition t that is mapped to a feature f by a mapping $m_{f,t}$, the following code needs to be appended to the effect of t for every mapped feature f:

> **if** $\neg f_c$ **then**
> $\quad f_c \leftarrow \top$

A test generator will set every control variable for all features associated with that transition, when this transition is added as a step to a test case. Hence, each control feature that is still *false* at the end of a test case indicates a free variation point. This result can be captured in a reusable test case for a subsequent selection of variants for testing.

3.2 An Evaluation of Product Line Test Generation

In this section, we present an implementation and evaluation of the two DC test design approaches described above, and compare them to AC test design methods.

3.2.1 A Prototypical Tool Chain

Our prototypical SPLTestbench is depicted in Fig. 13.

It consists of five major components:

(i) a feature injector to merge domain models as introduced in Section 3.1.2,

(ii) a model printer that exports the model to a test generator-specific format,

(iii) an adapter for third-party test generators,

(iv) a configuration extractor that collects incomplete configurations from the generated test cases, and

(v) domain-specific languages [20] that facilitate design and data processing of feature models, variability models, and configuration models.

An important component in this workbench is the model-based test generator (iii). This is a tool which takes a test model (e.g., a UML state machine), and produces a set of test cases from it. Since this chapter deals with test generation for product lines, we do not describe and compare different algorithms for model-based test generation. The interested reader is referred to the literature [21]. There are several test generators available, both from industrial and academic sources [22]. In our implementation and experiments, we used Conformiq Designer [23] and Real-Time Tester [24] as external test case generators.

Fig. 13 depicts the work flow a test engineer has to follow when using our SPLTestbench. The SPLTestbench is implemented as Eclipse plug-ins. The plug-ins are shown in the screenshot in Fig. 14. On the left-hand side the project layout is shown. In the center view, a currently opened UML state machine is depicted, which subsumes all possible behaviors of the SUT. On the right-hand side, we see the feature model in a tree view editor. Below, we find the feature mapping editor with a property view on the currently selected feature mapping. Fig. 15 shows how the components (i), (ii), and (iv) integrate into the Eclipse IDE. Each of the three menu items starts an individual wizard that guides the user through the details of the respective process.

3.2.2 Experimental Results

Now, we describe the experiments we performed to evaluate our approaches. We used several example product lines, including the eShop example from above. For our experiments we generated tests according to both presented approaches, AC and DC test design. For AC test design we sampled products according to two different feature model coverage criteria: *all-features-included-excluded* and *all-feature-pairs* [25]. As the name

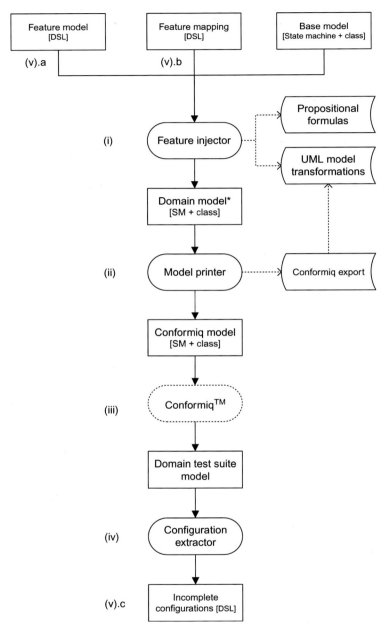

Fig. 13 Workflow of the SPLTestbench.

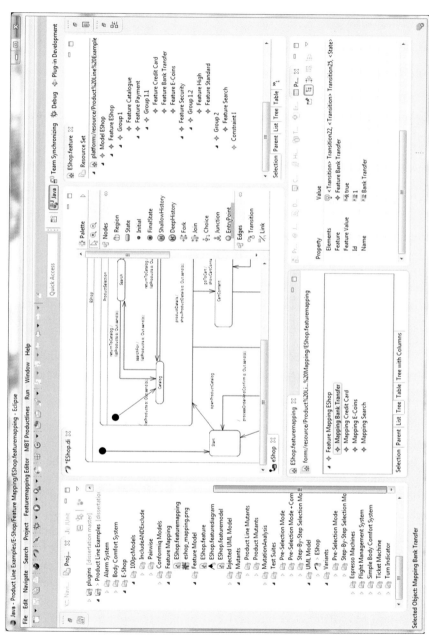

Fig. 14 Perspective of the SPLTestbench.

Fig. 15 Menu options in the SPLTestbench.

indicates, the criterion *all-features-included-excluded* is satisfied if in the set of sampled products, for each feature there is one product in which it is selected, and another one in which it is deselected. The criterion *all-feature-pairs* or *pairwise* holds of a sampling, if for every pair of features (f_1, f_2) there are products where f_1 and f_2 are both selected, one where neither f_1 nor f_2 is selected, one where f_1 is selected and f_2 deselected, and one where f_1 is deselected and f_2 selected. Details of the model sampling procedure will be described in Section 5.

Conformiq designer supports control-flow, branching, requirements, and path coverage criteria. For the individual state machine models as well as for the merged state machine model we required all-transitions coverage from the test generator [26]. There would be other, more sophisticated metrics for state machines to consider [27–29]. However, with this criterion we maintain comparability to other studies, e.g., in Ref. [30].

We were able to generate test suites for both approaches with all the aforementioned parameters for all examples. In order to compare the results, we counted

• the number of test cases,
• the test steps that were generated by the test generator, and
• the number of configurations that are necessary to execute the test cases.

These are important criteria when estimating test efforts: the number of configurations is a major factor since every configuration must be built, setup, and maintained for testing. The number of test steps is an indicator for the efforts for test design and maintenance. Finally, the number of tests is put in relation to the test steps. This gives the average test case length, which affects the efforts for debugging. With increasing length of a test case it becomes more difficult to isolate a fault within the SUT.

The results for these measures are shown in Table 1 for each individual approach: AC with all-features-included-excluded (AC-IX) as well as with pair-wise (AC-PW) coverage and DC with preconfiguration (DC-Pre) as well as with step-by-step (DC-Step). The AC-PW approach scores the highest values for all measures since it applies the strongest feature coverage

Table 1 Test Cases, Test Steps, and Configurations for Each of the Presented Approaches With the eShop Example

	Approach			
	AC-IX	AC-PW	DC-Pre	DC-Step
Tests	20	71	13	13
Test Steps	135	486	48	39
Configurations	2	7	4	2

criterion and thus covers a maximum of configurations. Consequently, more test cases and test steps are generated than for any other approach. In contrast, the DC-Step yields the lowest scores for any measure, while at the same time—as stated in Section 3.1.2—it is focused on covering every reachable transition. We take this as an indicator for DC test design to scale better than AC approaches. We performed similar experiments on several other examples, with comparable results.

Concluding, DC test design produces test suites with a significantly lower number of tests and test steps than AC test design. Thus, test execution efforts are much lower for these test suites. As we will see in Section 5, this does not necessarily lead to a lower error detection capability.

4. ASSESSMENT OF PRODUCT LINE TEST SUITES

In the previous section, we evaluated test generation algorithms with respect to the size and complexity of test suites. However, testing usually is performed to *detect errors* in an implementation. Thus, the complexity may not be the only quality criterion which is important for a test suite. The "effectiveness" of a test suite is often called its *error detection capability*. Given a test suite and an SUT, test execution will show up whether the implementation contains errors or not. Thus, by repeatedly executing a test suite with a faulty SUT, the error detection capability of the test suite can be measured. However, if the SUT does not exist yet, what is the error detection capability of the test suite? In a product line development, often it is impossible to construct all possible products in advance. As an example, the software for the embedded control units in a car is highly dependent on which features are present in the car. It is constructed only when the car is actually built, according to the orders of the customer. Given a test generation algorithm, how can we decide whether the generated test suite is adequate? Can we compare different product line test generation methods with respect to the quality of their result? Test assessment is an integral part for evaluation of the concepts presented in the previous section. This section builds the foundations for assessing the quality of a product line test suite in terms of its error detection capability. Though there are many methods proposed for testing a product line, until now, quality assessment of test suites mostly is limited to measuring code, model, and/or requirements coverage [31, 32].

A generally acknowledged approach to measure the error detection capability of a test suite is *mutation analysis*. Artificial faults are inserted into an SUT to form a *mutant*. Then, the test suite is executed and it is observed how many of the errors in the mutant are detected. In product line engineering, mutation analysis so far has been applied to individual products of a product line only. This "product-level approach" has two major drawbacks:

- First, developers can make errors in all kinds of artifacts, not only on product models. For example, there could be faults in the base model or variability model as well. This can lead to new kinds of faults in the products.
- Second, in the product-level approach the selection of products to mutate is biased by the selection of products for testing. Therefore, the mutation analysis assesses the quality of the tests for particular products. It is unclear how to combine the results to get an assessment for the whole product line.

To better understand which errors can occur in model-based product line engineering, we consider the different design paradigms discussed in Section 2 (annotative, compositional, and transformational variability modeling, see Fig. 3). We analyze potential errors in the respective design processes. From this, we develop mutation operators for variability and base models to mimic possible faults in these models.

To tackle the second problem, we define a mutation system and specific mutation operators for domain-level artifacts. This enables us to assess the test quality independently from the tested products. Subsequently, the error detection capabilities of a test suite can be assessed for the complete product line.

4.1 Potential Errors in Model-Based Product Line Engineering

The feature mapping has a major impact on the resolved products in a product line. However, designing a variability model is a complex and error-prone task, which is hard to automate. We identify potential errors in this process in a systematic way. We check each modeling paradigm for possibilities to

- omit necessary elements,
- add superfluous element, or
- change the value of an element's attribute.

For each potential error we discuss its effects onto the resolved products.

4.1.1 Annotative Variability Modeling

In the annotative variability paradigm, the following model elements for potential errors in the variability model can be identified: mappings, their attribute feature value, mapped feature, and the set of mapped elements. The errors which can be made on these model elements and their effects are as follows:

1. *Omitted mapping:* a necessary mapping is left out by its entirety. Subsequently, mapped elements will be part of every product unless they are restricted by other features. As a result, some or all products unrelated to the particular feature will include superfluous behavior. Products including the mapped feature are not affected by this error, since the behavior is now part of the common core and, thus, enabled by default.

2. *Superfluous mapping:* a superfluous mapping is added such that a previously unmapped feature is now mapped to some base model elements. This may also include adding a mapping for an already mapped feature, but with inverted feature value. Adding a mapping with feature value set to *true* results in the removal of elements from products unrelated to the mapped feature. Contrary, adding a mapping with feature value set to *false* removes elements from any product which the mapped feature is part of. In any case the behavior of at least some products is reduced.

3. *Omitting a mapped element:* a mapped model element is missing from the set of mapped elements in a mapping. Subsequently, a previously mapped element will not only be available in products which the said feature is part of, but also in products unrelated to this feature. As a result, some products offer more behavior than they should or contain unreachable model elements.

4. *Superfluously mapped element:* an element is mapped although it should not be related to the feature it is currently mapped to. As a result the element becomes unavailable in products which do not include the associated feature. The product's behavior is hence reduced.

5. *Swapped feature:* the associated features of two mappings are mutually exchanged. Subsequently, behavior is exchanged among the two features and thus, affected products offer different behavior than expected.

The result is the same as exchanging all mapped elements among two mappings.

6. *Inverted feature status:* the binary value of the feature value attribute is flipped. The mapped elements of the affected mapping become available to products where they should not be available. At the same time, the elements become unavailable in products where they should be. For example, if the feature value is *true* and is switched to *false*, the elements become unavailable in products with the associated feature and available to any product not including the said feature.

This enumeration covers all basic errors which can be made in the variability model. Other errors can be described by combinations of these basic errors. In the next section, we will use them to define fault injection operators.

4.1.2 Compositional Variability Modeling

In domain modeling with compositional variability, a mapping is a bijection between features and modules composed from domain elements. Potential errors in variability models can be made at mappings, mapped feature, and mapped module. Similar to above, we identify the following potential errors:

1. *Omitted mapping:* a necessary mapping is missing in its entirety.
2. *Superfluous mapping:* a superfluous mapping is added.
3. *Swapped modules:* the associated modules of two mappings are mutually exchanged.
4. *Swapped features:* the associated features of two mappings are mutually exchanged.

4.1.3 Transformational Variability Modeling

For other paradigms, like delta modeling [12], we make similar observations. In contrast to compositional variability models, transformational variability models start from an actual core product, instead of a base model. From this, only the differences from one product to another are defined by delta modules.

Similar to above, errors in delta modeling include omitted or superfluous deltas, omitted or superfluous base elements, and omitted or superfluous delta elements. Since there are no attributes in deltas other than add/remove, changes in deltas can be neglected.

To sum up, a systematic analysis according to the three categories 'add," "remove," and "change" yields a list of all potential errors for each modeling paradigm. Subsequently, we will use the identified error possibilities in a fault injection framework to assess product line test suites.

4.2 Mutating Domain Models

As discussed in the previous subsection, model-based product line engineering bears the risk for new kinds of errors compared to single systems engineering. Current test design methods and coverage criteria are not prepared to deal with these errors and resulting faults. We propose a mutation system for assessing the error detection capability of a product line test suite. Note that in contrast to other authors, we assess the test quality for the whole product line rather than for single products.

To define our mutation system we need novel mutation operators. Mutation operators defined for nonvariant systems cannot add or remove features. However, a high-quality test suite should also detect such faults. Hence, we propose new mutation operators based on the potential errors identified in Section 4.1.

4.2.1 A Mutation System for Product Lines

Performing mutation analysis on product line tests is different from nonvariant system tests, since in contrast to conventional mutation systems, a mutated domain model is not executable per se. Thus, testing cannot be performed until a decision is made toward a set of products for testing. This decision depends on the product line test suite itself, since each test is applicable to just a subset of products.

In Fig. 16, we depict a mutation process for assessing product line test suites, which addresses this issue. Independently from each other, we gain (a) a set of domain mutants by applying mutation operators to the domain model and identify (b) a set of configurations describing the applicable products for testing. We apply every configuration in (b) to every mutant in (a), which returns a new set of product model mutants. Any mutant structurally equivalent to the original product model is immediately removed and does not participate in the scoring. The mutant product models are resolved to mutant products, and, finally, tests are executed. Our mutation scores are based on the domain model mutants; hence, we establish bidirectional traceability from any mutant domain model to all its associated product mutants. If a product mutant is detected (killed) by a test, we backtrack its domain model mutant and flag it accordingly. The mutation score is the quotient of the detected and overall number of mutants.

In the mutation or resolution process, models may be generated which are invalid (syntactically incorrect). Of course, such mutants must be discarded. For all mutation operators, there is a risk that mutants are materialized which behave equivalent to the original product (so-called masked

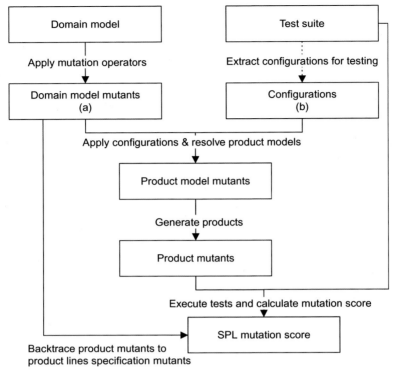

Fig. 16 Mutation process for product line systems.

mutants). For example, a masked mutant is generated when the mutated element is part of a disabled feature. Of course, masked mutants should be excluded from the scoring. However, the question whether a mutant is masked is in general undecidable. Our mutation systems filters mutants which are structurally equivalent to the original product model.

4.2.2 Product Line Mutation Operators

Here, we present a list of mutation operators for variability models with annotative variability. We start with standard state machine mutation operators, applied to the domain level.

Mutation operators for the base model: In general, for product line test assessment it is not necessary to mutate classes in UML state machines [33]. The system's logic is designed in the state machine diagrams, while the classes are merely containers for variables and diagrams. Many mutation operators for state/transition systems have been defined in the literature, see, e.g., Refs.

[34–37]. For product line tests, it is sufficient to consider only operators based on transitions, as these have the strongest impact on the behavior of the SUT.

Our mutation system provides the following mutation operators for the base model:

- Delete transition (DTR)
- Change transition target (CTT)
- Delete effect (DEF)
- Delete trigger (DTI)
- Insert trigger (ITG)
- Delete guard (DGD)
- Change guard (CGD)

These operators are selected and implemented such that they cover most of the state machine mutation operators which are found in the literature.

Mutation operators for the variability model: Now, we describe mutation operators according to the potential errors in the feature mapping identified in Section 4.1. We comment on hidden, invalid, and equivalent mutants where appropriate.

Delete mapping (DMP): The deletion of mapping will permanently enable the mapped elements, if they are not associated to other features that constrain their enabledness otherwise. In our examples, no invalid mutants were created. However, for product lines that make heavy use of mutual exclusion (Xor and excludes) this does not apply. The reason for this are competing UML elements like transitions that would otherwise never be part of the same product. Multiple enabled and otherwise excluding transitions are possibly introducing nondeterminism or at least unexpected behavior.

Some product mutants created with this operator might behave equivalent to an original product. This is the case for all products that include the feature for which the mapping was deleted. Since these mutants are structurally equivalent to the original product model, they are easy to detect.

Delete mapped element (DME): This operator deletes a UML element reference from a mapping in the variability model. It resembles the case, where a modeler forgot to map a UML element that should have been mapped.

Similar to the delete mapping operator, this operator may yield nondeterministic models, where otherwise excluding transitions are concurrently enabled. Product mutants equivalent to the original product model can be derived, if the feature associated to the deleted UML reference is part of the product. Again, this results in structural equivalence to the original product.

Insert mapped element (IME): This operator inserts a new UML element reference to the mapping. This is the contrary case to the operators defined before, where mappings and UML elements were removed. However, inserting additional elements is more difficult than deleting them, since a heuristic must be provided for creating such an additional element. We decided to copy the first UML element reference from the subsequent mapping. If there are no more mappings, we take the first mapping. This operator is not applicable if there is just one mapping in the feature mapping model.

Again, there is a chance of creating invalid mutants: If a UML element reference is copied from a mutually excluded mapping, the resulting model may be invalid due to nondeterminism.

Also structurally equivalent mutants are created when the features from the subsequent mapping, which acts as source for the copied element, and the target mapping are simultaneously activated in a product.

Swap feature (SWP): Swapping feature exchanges the mapped behavior among each other. This operator substitutes a mapping's feature by the following mapping's feature and vice versa. The last feature to swap is exchanged with the very first of the model.

Nondeterministic behavior and thus invalid models may be designed by this operator. This is due to the fact that the mutation operator may exchange a feature from a group of mutually exclusive features by an unrestricted feature. In consequence, the previously restricted feature is now independent, while the unrestricted feature joins the mutual exclusive group. This may concurrently enable transitions which results in nondeterministic behavior.

We gain structurally equivalent mutants if the two swapped features are simultaneously activated.

Change feature value (CFV): This operator flips the feature value of a mapping. A modeler may have selected the wrong value for this Boolean property of each mapping.

The operator must not be applied to a mapping, if there is a second mapping with the same feature, but different feature value. Otherwise, there will be two mappings for the same feature with the same feature value, which is not allowed for our feature mapping models.

This operator may yield invalid mutants, if it is applied to a mapping that excludes another feature. In that case, two otherwise excluding

UML elements can be present at the same time, which may result in invalid models, e.g., two default values assigned to a single variable or concurrently enabled transitions.

There are other possible domain model mutation operators, which, however, are of minor importance. For example, inserting superfluous mappings does not seem to be necessary: it remains unclear which and how many UML elements should be selected for the mapping. In most cases, such an operation will lead to invalid mutants.

In a basic mutation experiment, each operator is used for each applicable model element exactly once. To get an idea of the complexity, consider the following numbers: For our eShop example, which has about 10 features and 44 state chart elements (states and transitions), after filtering this yields 152 product line mutants, which are resolved into 574 mutated products. Of course, it is possible to generate an arbitrarily higher amount of mutants by repeated application and combination of mutation operators. However, it is doubtful whether this brings about new insights.

4.3 Evaluation

To evaluate the above ideas, we performed a mutation analysis on three example product lines, including our eShop example. With our prototypical SPLTestbench, we developed and adapted the domain models for the examples. Then we derived a test suite for each example by applying the techniques described in Section 3. The external test generator was set to transitions coverage. For the eShop example, this generated 13 test cases with a total of 103 test steps. From the generated test suite, the SPLTestbench selected 4 variants for testing and resolved mutated product models from the mutated domain model.

Since there were no implementations available for our examples, we decided to generate code from the mutated product models and run the tests against them. (Note: Generating code *and* tests from the same basis for testing the code is not advisable in productive environments, since errors propagate from the basis to code *and* tests. However in our case, tests are executed against code derived from mutated artifacts, while the test cases were generated from the original model.) To derive our implementations, we implemented a small code generator transforming product models into Java. Another transformator generates executable JUnit code from the tests which we gained from the test generator. The mutation system then collects all the code artifacts, executes the tests against the product code, and finally reports the mutation scores for all tests and for every operator individually.

For the eShop example, on the 574 mutated products there were in total 1855 tests executed. All of the transformations above and the mutation system are integrated into the SPLTestbench. The test execution is fully automatic and is done within a few minutes.

For each mutation operator we measured the mutation score as the quotient of detected mutants by generated mutants. The results for all mutation operators from the previous subsection is shown in Table 2. In the last column of this table, we show the weighted average for each mutation operator. The weighted average takes the number of product line mutants for each example and operator into consideration.

The table shows some interesting effects. For most of the operators we gain mutation scores between 60% and 100%. This is in the expected range for test suites generated with the all-transitions coverage on random mutants [30, 38]. However, there are some notable exceptions: DMP (0%) and DME (0%) on feature mappings, ITG (21.67%) and DGD (41.18%) on base models. The reason is that these operators add superfluous behavior to the product. Errors consisting of additional (unspecified) behavior are notoriously hard to find by specification-based testing methods. Specification-based testing checks whether the specified behavior is implemented; it cannot find unspecified program behaviors (e.g., viruses). Program code-based testing

Table 2 Mutation Scores Per Operator (and Number of Products)

Op.	eShop	Example #2	Example #3	Average
DMP	0% (4)	0% (5)	0% (8)	0%
DME	0% (14)	0% (8)	0% (21)	0%
IME	75% (4)	40% (5)	50% (8)	53%
SWP	100% (4)	60% (5)	63% (8)	71%
CFV	100% (4)	100% (5)	87.50 (8)	94%
DTR	89% (28)	84% (19)	63% (19)	80%
CTT	64% (28)	63% (19)	37% (19)	56%
DEF	100% (16)	82% (17)	62% (13)	83%
DTI	83% (23)	100% (13)	94% (17)	91%
ITG	21% (24)	28% (18)	17% (18)	22%
DGD	0% (1)	43% (14)	50% (2)	41%
CGD	100% (2)	69% (48)	90% (10)	73%

checks whether the implemented behavior is correct with respect to the specification; it cannot find unimplemented requirements (e.g., missing features).

To solve this problem, we must not only check whether a required feature is implemented, but also whether a deselected feature is really absent. One possibility for this is to add *product boundary tests*, which are automatically generated from the domain model. A detailed elaboration of this idea can be found in Ref. [39]. Another possibility is to admit so-called negative tests, which are constructed manually and test whether a certain behavior is absent. This idea will be followed in Section 6.

5. TEST-DRIVEN PRODUCT SAMPLING

In a large product line with one hundred or more features, there might be up to $2^{100} \simeq 10^{30}$ possible products. Clearly, it is impossible to build and test all of these products. In this section, we address the question which products of a large product line should be selected for testing. In practical applications, the choice is often done by some heuristics (e.g., "test one minimal and one maximal product with respect to the number of features," or "test those products which the customers are most likely to order").

However, it is not clear whether the error detection capabilities of these heuristic choices are acceptable. From a systematic point of view, there are different criteria on the selection. One possibility is to select a minimal number of different configurations such that each test case is executable in at least one product. This minimizes the amount of tested products, and thus reduces the testing effort. Other possibilities are to minimize or maximize the size of product configurations in terms of activated features, or the diversity of configurations. The question is which method offers an acceptable error detection rate in relation to the testing effort.

Subsequently, we address the question how the sampling of configurations from test cases affects the effectivity of product line testing. Using our domain-centered test design methods described in Section 3.1, the domain test cases can be used to sample product configurations for testing. We describe experiments to measure the effect of different sampling criteria on the error detection rate. We consider the following criteria:
- Sampling as many or as few configurations as possible,
- sampling large or small products in terms of activated features, and
- sampling diverse or random products.

Our expectation is that selecting many, large, and diverse products yields the highest fault detection capability and test effort. First, testing many products should decrease the chance of missing faults which are specific to some particular combination of features. Second, selecting large products for testing should expose faults arising from feature interactions. Third, testing diverse products, rather than testing similar products, should increase the error detection capability.

In order to confirm these expectations, we develop a method for product selection from domain test cases. This method allows us to perform experiments with different sampling criteria. The assessment of error detection capabilities is done by the mutation system presented in the previous section. We analyze the results for all three examples of the previous section, including the eShop, and give recommendations for practical use.

From the resulting product configurations, products can be resolved and finally the tests can be executed against their associated product. In the next section, we present optimization criteria for sampling configurations from such generic test cases.

5.1 Sampling Configurations From Generic Test Cases

In this section we address the question how to sample product configurations from a test suite. In domain-centered test design, an incomplete product configuration is created and stored with each test case during test generation. Product configurations can be sampled from these incomplete configurations by assigning a concrete value (*true* of *false*) to each undecided feature. Since there is a choice to make, we can apply an optimization criterion. The sampling should be such that the likelihood of detecting faults in the product line should be maximal, while the test effort is kept reasonably low. Since test cases were selected according to a model coverage criterion, every test case should be executable at least once. Moreover, in order to keep test efforts low, test cases should not be executed more than once.

Due to the fact that feature models are representable as propositional formulas, the problem of sampling configurations can be viewed as a Boolean satisfiability problem. In order to search for an solution to an optimization criterion, we represent it as a constraint satisfaction problem. First, we describe the problem of sampling a product configuration from an incomplete configuration as a propositional formula. On this basis, we then define minimization and maximization criteria to sample large, small, few, many, and diverse variants from a set of incomplete configurations.

5.1.1 Random Sampling

The first challenge is to complete a given incomplete product configuration. This problem can be solved straight-forward by an SAT solver. We use the Java constraint programming solver JaCoP [40] which is based on the DPLL SAT algorithm. We declare for each feature f in F a new variable v_f with an appropriate domain. The domain varies depending on the feature's assignment:

- $f = true$ then the domain of v_f is $\{1\}$
- $f = undecided$ then the domain of v_f is $\{0, 1\}$
- $f = false$ then the domain of v_f is $\{0\}$

An additional constraint is the propositional formula representing the feature model (see Fig. 2). JaCoP is now able to make a random assignment to each undecided feature and check the solution for consistency with the feature model. This method can be repeated individually for every incomplete configuration in the given test suite.

5.1.2 Targeted Sampling

The above procedure yields a set of product configurations which may not be optimal. From each test case, a separate product is sampled. There is no systematic check whether a product is appropriate for several test cases. In the following, we define and use optimization criteria for this purpose. We consider the following criteria:

- Few/many configurations,
- small/large configurations (in terms of selected features),
- diverse/random configurations, and
- combinations of these criteria.

The optimization should be such that for every test case there is a valid configuration, and for every sampled configuration there is an associated test case. Thus, for a test suite with m test cases, we will not sample more than m products. For an automated sampling of products, we now formulate these criteria as Boolean optimization problems, which can be solved automatically by a constraint solver.

5.1.2.1 Optimizing the Amount of Distinct Configurations

The aim is to select either few or many products to execute every test case in the given test suite exactly once. From a product configuration with features $F_n : f_1, \ldots, f_k$, in JaCoP we define the binary number

$$b_n = (v_{f_1} v_{f_2} \ldots v_{f_{k-1}} v_{f_k})_2,$$

where v_{f_i} is the variable associated with feature f_i (see above). For a test suite with m test cases, we define the set Z as the collection of all b_i, where $1 \leq n \leq m$:

$$Z = \{b_1, b_2, \ldots, b_{m-1}, b_m\}$$

With such an encoding, we can ask the constraint solver to find a variable assignment for the v_{f_i} respecting additional criteria. To optimize the amount of distinct configurations, we define a cost function as the cardinality of Z:

$$cost_a = |Z|$$

Now we can ask JaCoP to find a solution which minimizes or maximizes this cost function. The resulting set of configurations has between 1 and m elements.

5.1.2.2 Optimizing the Size of All Configurations

For minimizing the size of a set of configurations, we define its size as the sum of all selected features in all configurations. For constraint solving, we interpret "selected'" as numerical value "1" and "deselected" as "0," respectively. In JaCoP, we define the size of a single product configuration as follows:

$$s_n = \sum_{i=1}^{k} v_{f_i}.$$

When we accumulate sizes of all product configurations, we can ask JaCoP to optimize toward either a minimal or a maximal overall size:

$$cost_b = \sum_{n=1}^{m} s_n.$$

Here, maximization achieves large product configurations and minimization, small product configurations. The cost of the smallest solution is $2 \times m$, having the root feature and only one other feature enabled (2), multiplied by the amount of test cases (m). The highest cost is $k \times m$, where every feature k is selected in every test case m. This is the result of assigning a single product with all features activated to all test cases.

5.1.2.3 Optimizing the Diversity of Configurations

Similar to Section 5.1.2.1 and 5.1.2.2, we define diversity over a set of m test cases and k features. First we establish a relation between a single feature i

over all configurations. The goal is to have each feature as often selected or deselected to obtain the most different assignments.

We achieve this by calculating the diversity d_i of each feature $f_{n,i}$, where $1 \leq n \leq m$ and $1 \leq i \leq k$:

$$d_i = \sum_{n=1}^{m} v_{f_{n,i}}$$

Next, we calculate the deviation from optimal diversity, which is $m/2$, because we want a feature to be equally often selected and deselected over all n configurations. Subsequently, the deviation of a feature f_i from its optimal diversity is calculated by $|d_i - (m/2)|$. Finally, we achieve maximal diversity by minimizing the sum of all deviations:

$$\text{cost}_c = \sum_{i=1}^{k} |d_i - (m/2)|$$

The minimal cost for a solution to this problem is 0 with product configurations being maximally diversified. The highest cost is $(m/2) \times k$, where the same configuration is sampled for every test case.

It should be noted that this approach does not maximize the amount of sampled product configurations, but their diversity. Inherently, this leads to a solution with few unique product configurations, although there might be solutions with more product configurations but less diversity. A possibility to increase the amount of product configurations is to combine the two criteria diversity and maximization.

5.1.2.4 Combinations

Now we look at combinations of the previously defined constraints, e.g., many with large and diverse configurations. All the above criteria can be combined, with the exception of

- few with many product configurations, and
- small with large product configurations.

Of course, costs cannot be summed up directly if the optimization targets are conflictive, e.g., if large and diverse should be combined, the targets are minimization and maximization. In this case, a decision for an overall optimization target must be made (min *or* max) and the costs of the criterion not fitting that target must be inverted. Costs are inverted by subtracting the solution's costs from the expected maximal costs.

5.2 Evaluation

For evaluation of the proposed methods, we extended our SPLTestbench to support optimization-driven product sampling from incomplete product configurations. The sampling process is supported by Eclipse plug-ins to configure the sampling and start the sampling process as depicted in Fig. 17.

As an example to evaluate the optimization criteria, we use our eShop product line. Test case generation from the variability model was discussed in Section 3.1. Here, we use the test cases generated with the step-by-step method to measure the effects of sampling for different optimization criteria on the test case's error detection capability. We compare the results to include/exclude-all-features and pair-wise combinatorial testing from application-centered test design.

For assessing the error detection capability, we use the product line mutation framework presented in Section 4.2.1. For our experiment, we apply both of the supported types of mutation operators: behavioral operators, which mutate the state machine model, and variability operators, which mutate the variability model. However, we did not apply the operators DMP, DME, and ITG. As discussed before, mostly these operators add superfluous behavior, which cannot be detected by model-generated test suites.

We sampled configurations for all sampling criteria in isolation and for the four combinations Few+Small+Div, Few+Large+Div, Many+Small +Div, and Many+Large+Div. As a comparison value, we also consider

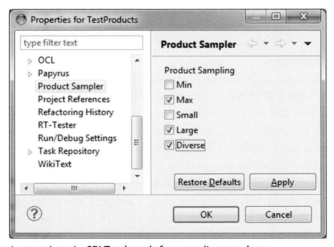

Fig. 17 Project options in SPLTestbench for sampling products.

Table 3 Results of Product Sampling

Isolated DC Samplings	Configurations	Executed Test Steps	Mutation Score Base Model	Mutation Score Variability Model
Few	1	48	78%	33%
Many	13	48	77%	92%
Small	4	48	77%	92%
Large	1	48	78%	33%
Div	7	48	78%	67%
Combined DC samplings				
Few + Small + Div	4	48	77%	92%
Few + Large + Div	4	48	78%	67%
Many + Small + Div	10	48	77%	92%
Many + Large + Div	10	48	78%	67%
AC samplings				
Include/exclude-all	2	135	77%	67%
Two-pair	7	486	78%	100%

the numbers for AC test design. The results can be seen in Table 3. In this table, the number of automatically sampled configurations is shown in the first column. This number is responsible for much of the testing effort. The second column shows the number of test executions. For a given DC sampling, every test case is executed exactly once. In contrast, for AC test design, test cases are generated individually for each variant. Hence, here we gain more test cases and subsequently more test executions. The third and fourth columns give the achieved mutation score on the base model and on the variability model, respectively.

The test suite contains 13 test cases with 48 test steps. Since this test suite is the same for all samplings, the procedure indeed assesses the impact of the different sampling criteria on the test suite's error detection capability. As can be seen from the data, the error detection capability of the test suite varies with the applied sampling criterion. The mutation scores for the base model exposes only marginal variance (76.5%–77.6%). This is due to our test design approach where all-transitions coverage as test selection criterion is applied to base model and product models.

More variance is found in the error detection capability for errors in the variability model: the highest error detection capability of isolated criteria is achieved by maximized amount of products (Many) and small products (Small). In contradiction to our expectations, sampling large products reduces the error detection capability. This may be due to the fact that only few products are being sampled and tested. Though the criteria "Many" and "Small" have a comparable error detection capability (91.7%), "Small" is more efficient in terms of testing effort, since more products have to be provided.

For combined sampling criteria the variance of the mutation scores declines (66.7%–91.7%). The highest scores in the group of combined criteria are achieved by the following two combinations: maximized amount with small and diverse products (Many + Small + Div) and minimized amount with small and diverse products (Few + Small + Div). For these, efficiency is higher for Few + Small + Div than for Many + Small + Div, since less products are sampled for testing. In general, combined criteria scored equal or lower to the top scoring isolated criteria, but never better.

The DC scores lie in between the two presented AC criteria. However, AC testing needs much more resources: For a comparable score, two-pair testing needs ten times more test steps than any DC test.

Summing up, application-centered and domain-centered testing differs only if errors in the variability model are of concern. Within different DC methods, sampling few, small, and diverse products seems to be most advantageous. We performed a similar analysis for the two other examples mentioned in the previous section, with comparable results.

6. ASSIGNMENT OF PRODUCT LINE TEST CASES

In Section 3, we considered different ways of generating test suites from domain models. Subsequently, we used these tests for product sampling and mutation analysis. Thus, the domain model was primarily used for test generation. In model-based engineering, either products or tests can be derived from a given model. In this section, we consider the setting that domain models are used for deriving implementations, whereas test cases are developed separately and manually. This setting describes the actual situation in most medium and large companies, where testing and implementation is done in different departments. In fact, in this section we do not care about the test case development process; we just assume that there is a significant body of test cases available, and that these test cases were developed independently from the products. This situation is found in all companies transitioning to a model-based software engineering process, which already have a large number of legacy test cases.

For such a setting, the question is, which test cases should be executed with which products? Certainly, there are syntactic criteria excluding a test case to be applicable for a certain product. For example, a test case might address an interface which is not present in the product. In such a case, the test cannot be executed with this product.

However, even if all syntactic criteria are being met, a test case might not be relevant for a certain product. For example, the specification may not determine whether the behavior tested by t_1 is required or not for a product p. That is, p may or may not show the behavior which is tested by t_1. In such a case, no new information about the conformance of the product would be obtained by executing the test case. Thus, it does not make sense to execute t_1 with the SUT p.

As another example, consider a test case t_2 for an advanced feature which is not present in a basic product p_b. Here, we would expect the test execution of t_2 with SUT p_b to fail. It may also be advantageous to execute such a *negative test case* in order to check that the basic product has no superfluous behavior.

Closely following ideas presented in Ref. [41], in this section we show how the product model can be used to determine the expected outcome of a test case to a product. The exposition is based on our previous work [42], where we give formal definitions, an extended example, and a detailed description of the algorithm.

6.1 Colored Test Cases

We use a three-valued coloring scheme to capture what design decisions have been made in the domain model and configuration with regard to a specific product: A test case is colored green if it reflects a behavior that is expected from this particular variant of a software product line. It is colored red if the variant should not allow the described behavior. Finally, a test case is colored yellow if the respective behavior is neither required nor disallowed by the specification of the variant. This can happen, e.g., if the specification is nondeterministic or incomplete.

Intuitively, green test cases reflect *required* and red test cases *forbidden* behavioral properties of the specification. Yellow tests mirror *open design decisions*, i.e., properties which are not (yet) decided in the specification. Since the color of a test case depends on the base model as well as the variability model and its resolution for a particular variant, the same test case can be green for one product, but red or yellow for another one.

The three-valued scheme is similar to the definition of an incomplete configuration in domain-centered test design (Section 3.1.2). The difference is that an incomplete configuration determines for each test case the set of products for which it is applicable, whereas the coloring determines for each test case and product whether it is a required, allowed, or forbidden behavior.

In order to make these notions more precise, we briefly recall the UML stipulations on the execution of a model: In UML state machines, a transition $e[g]/a$ may have a *trigger e*, can be restricted by a *constraint g*, and can invoke a *behavior a*. The UML superstructure explains: "A trigger specifies an event that may cause the execution of an associated behavior. An event is often ultimately caused by the execution of an action, but need not be. [...] Upon their occurrence, events are placed into the input pool of the object where they occurred [...]. An event is dispatched when it is taken from the input pool and is processed by the classifier. At this point, the event is considered consumed and referred to as the current event." [43, p. 471sq.]. The constraint language is not specified in UML; "a constraint is a condition or restriction expressed in natural language text or in a machine readable language for the purpose of declaring some of the semantics of an element" [43, p. 57]. A behavior is a consequence of the execution of an action by some related object. The behavior invoked as the effect of a transition may contain several actions, e.g., calling an operation, changing variable values, or causing the occurrence of some event.

To define the notion of a test case, we fix a *test signature* Σ. In our approach, we assume that Σ is a subset of the occurrences and dispatches of events which are contained in the product model. In this case, we say that the test case is applicable to the product model.

Additionally, we require that stimuli can be sent to the SUT, for example, pressing a button, from the outside. We represent this as the artificial entity *tester*. Events sent from the tester are called *input* events for the SUT, other events are called *output* events for the SUT. Intuitively, elements of the signature are the only events which can be "noticed" by the test case; events not in the signature are "invisible." A *test case* formally is a finite sequence of elements from the test signature Σ.

In order to fix the color of a test case, we assume that there exists a function *enabled* assigning to each configuration of a UML model the set of elements from Σ which may occur next. That is, an event $e \in \Sigma$ is in *enabled*(c), if upon its occurrence there is a sequence $c_0 \xrightarrow{e_1} c_1 \xrightarrow{e_2} \cdots \xrightarrow{e_n} c_n$ of transitions such that $c_0 = c$ and $e_n = e$, and for all $i < n$ it holds that $e_i \notin \Sigma$. In this case, we say that c_n is *reached* from c by e. A useful assumption, which is, however, not required for our theory, is that input events from the tester are enabled at any time. For an event $e \in$ *enabled*(c), we say that it is *testable* at c, if it is either an input event, or it is an output event and it is not the case that some output event $e' \in \Sigma$ different from e is enabled in c. Intuitively, if an output e is testable at c, it is the output from Σ which must occur next, if any.

Since UML contains semantic variation points, the function *enabled* is tool-dependent. In particular, UML does not impose an ordering on events in the event pool; furthermore, the mechanism for determining the behavior to be invoked as a result of a call operation is unspecified, and it is a semantic variation point whether one or more behaviors are triggered when an event satisfies multiple triggers. The UML allows an event to be dispatched in a configuration even if there is no transition taking this event as a trigger; in such a situation, this event is discarded.

The *color* of a test case $T = \langle e_1, \ldots, e_n \rangle$ in the signature Σ with respect to a product model is a value from {*green*, *red*, *yellow*}, such that

- *color*(T) = *green* if for all $k < n$ and every sequence $\langle c_0, c_1, \ldots, c_k \rangle$ of configurations such that c_0 is an initial configuration, and c_i is reached from c_{i-1} by e_i for all $1 \leq i \leq k$ it holds that e_{k+1} is testable at c_k;
- *color*(T) = *red* if there is no sequence $\langle c_0, c_1, \ldots, c_n \rangle$ of configurations such that c_0 is an initial configuration, and c_i is reached from c_{i-1} by e_i for all $1 \leq i \leq n$; and
- *color*(T) = *yellow*, otherwise.

In other words, a test case is green if it can be observed in all possible executions of the model triggered by this test case. It is red if there is no possible execution where it can be observed. It is yellow if some executions show the behavior and others do not.

Note that our definition enforces that for each test case $T = \langle e_1, \ldots, e_n \rangle$ for which *color*(T) = *green* there is at least one sequence $\langle c_0, c_1, \ldots, c_n \rangle$ such that c_0 is an initial configuration, and for all $1 \leq i \leq n$, configuration c_i is reached from c_{i-1} by e_i. That is, green test cases must indeed be observable in the system's executions.

For example, consider a minimal eShop where no search function is available. For such an eShop, the following test case would be green:

> *openProductCatalog* → *listProducts* → *productDetails* → *showProductDetails* →
> *addProductToCarts* → *proceedToCheckout* → *showSummary* → *selectPayment-*
> *Method* → *startPayment* → *selectBankAccount* → *proceedPayment* → *validate-*
> *Payment* → *paymentComplete* → *valid* → *processOrderAndConfirm.*

The following test case, which tests for the search feature to be absent, would be red:

> *openProductCatalog* → *listProducts* → *searchFor* → *listProducts.*

Here are some simple properties of our coloring.
- An empty test case (consisting of no events at all) is always green.
- A one-element test case is green if its event is enabled and testable in all initial configurations; it is red, if the event is initially not enabled; and yellow, if it is enabled in some initial configuration but not testable.
- Any initial fragment of a green test case is green; any extension of a red test case is red.
- If a state is nondeterministic, e.g., from state s there are transitions $/a$ and $/b$, then the test cases $\langle a \rangle$ and $\langle b \rangle$ are yellow, since $enabled(s) = \{a, b\}$, but a is not testable at s. Assuming that the test signature is $\{a, b, c\}$, the test case $\langle c \rangle$ is red, since neither $/a$ nor $/b$ produces c and thus c is not enabled in s.
- Consider a situation where the effect of a transition invokes a behavior expression including an operation for which only its signature is known (e.g., a transition $/obj.op(arg)$, where the operation op is declared in the class diagram, but the return value of op for a given argument arg is not specified). Then test cases using such a transition will be yellow, as all possible return values are enabled in the state machine; however, the test case contains only a specific one.

The test verdict (pass or fail) for a test is assigned by executing a green or red test case with a concrete product. A product passes a test suite, if it behaves as expected, i.e., if it exhibits the behavior described in all green test cases and deviates from the behavior described in all red test cases. Yellow test cases do not contribute to the detection of faults; thus we do not execute them.

6.2 Automated Test Coloring via Model Checking

For automating the above defined test coloring procedure for a given materialization of a product line and a test case, we use the tool HUGO, which is a UML model translator for model checking [44]. In particular, HUGO resolves the UML's semantic variation points mentioned above in a particular way thus also fixing the *enabled* function: the event pool is implemented as bounded event queues; since inheritance is not supported, the dispatching algorithm becomes straightforward, as no overloading has to be considered; only a single, nondeterministically chosen behavior can be triggered by a given event; and events which trigger no outgoing transition in a state configuration are silently consumed.

HUGO translates both the materialization and a test case over a test signature into Promela, which is the input language of the model checker SPIN [45]; syntactically, it is first ensured that the test signature indeed is a subset of the possible event occurrences and dispatches of the materialization. The resulting encoded product model shows instrumentation for observing all events: whenever an event occurs or is dispatched in the product model, a notification is sent out which can be used by an observer. The test case results in an automaton process sending those events to the system which occur at the *tester* and also reacting to those produced by the system: if an event of the test case is observed, the test case automaton proceeds; if any other event which is part of the test signature happens, the automaton goes to a dedicated *failure state*; events not present in the test signature are ignored. After successful observation of the last entry of the test case sequence the automaton enters a dedicated *final state*.

We have implemented the above translations and colored a number of test cases. The results often are surprising, the calculated color is different from the intuition. Fortunately, due to the counterexample mechanism of the model checker, we were able to analyze this discrepancy. It turned out that in all cases the intuition was mistaken; test cases which we assumed to be green (required) or red (forbidden) really were yellow (allowed but neither required nor disallowed). This shows the viability of our approach.

In this section, we have presented a theory and prototypical implementation for test case assessment in the model-based development of multivariant systems. To our knowledge, this is the first treatment of the subject in the context of UML-based software development.

We deal with both positive (green) and negative (red) test cases, and introduce a third color (yellow) for test cases whose outcome is not determined with a given product model. This means that it is needless to execute them with products based on this model. Our approach thus allows to assess and select those test cases from a universal test suite which are relevant for a given product. It would be a straightforward extension to extend our coloring to sets of products defined by incomplete configurations as defined above. Additionally, lifting our approach to logical, abstract test specifications in the universal test suite would be of interest. This would have to include different colorings for the different concretizations. For conciseness, we do not pursue these extensions further here.

Our test case coloring theory is well suited for testing deterministic reactive systems under test, where the response functionally depends on the provided stimuli. In the UML specification, it can deal with indeterminacy caused by semantic variation points and nondeterminism by underspecification and open design decisions, by assigning the respective test cases the color yellow. The theory excludes to formulate test cases for systems which are inherently nondeterministic. This can be the case, e.g., for a network of cooperating devices with unpredictable message delays. To deal with such a situation, we are investigating trees and UML interactions as test cases and the relation to the testing theory of de Nicola and Hennessy [46].

7. RELATED WORK

In this section, we give some references to related work. Due to the large volume of available literature on model-based testing and software product lines, this is necessarily only a small selection.

7.1 Specification of Software Product Lines

Software product families have already been proposed by Parnas in 1976 [1]. There are annual conferences dealing with product line engineering, e.g., the International Systems and Software Product Line Conference SPLC. An introductory textbook on this topic is by Pohl et al. [47].

A standardization attempt which incorporates many of the modeling concepts presented in this chapter has been the OMG initiative for CVL, the common variability language [48]. Unfortunately, due to patent disputes, this standardization could not be completed. Nevertheless, most of the concepts are now generally acknowledged and implemented by current product line engineering tools.

7.2 Model-Based Test Generation

Testing is one of the most important quality assurance techniques in industry. Since testing often consumes a high percentage of project budget, there are approaches to automate repeating activities like, e.g., regression tests. Some of these approaches are data-driven testing, keyword-driven testing, and model-based testing. There are many books that provide surveys of conventional standard testing [49–51] and model-based testing [21, 23, 52]. Modeling languages like the UML have been often used to create test models. For instance, Abdurazik and Offutt [35] automatically generate test cases from state machines.

Testing is an important topic also in the software product line literature. Systematic reviews can be found in Refs. [53–55]. These surveys show that much of the work on SPL testing is concerned with the question of selecting products for testing (see below).

7.3 Feature Modeling

Feature models are commonly used to describe the variation points in product line systems. Our feature modeling language is similar to other standard feature languages, e.g., Ref. [56]. There are several approaches to apply feature models in quality assurance. For instance, Olimpiew and Gomaa [57] deal with test generation from product line systems and sequence diagrams. In contrast to sequence diagrams, state machines are commonly used to describe a higher number of possible behaviors, which makes the combination with feature models more complex than combining feature models and sequence diagrams. As another example, McGregor [58] shows the importance of a well-defined software product line testing process. Pohl and Metzger [59] emphasize the preservation of variability in test artifacts of software product line testing. Lochau et al. [18] also focus on test design with feature models. In contrast to our work, they focus on defining and evaluating coverage criteria that can be applied to feature models.

7.4 Mutation Analysis

Mutation analysis for behavioral system models, e.g., finite state machines, is a well-established field. Fabbri et al. introduced mutation operators for finite state machines in Ref. [34]. Belli and Hollmann provide mutation operators for multiple formalisms: directed graphs, event sequence graphs [36], finite state machines [35], and basic state charts [37]. They conclude that there are two basic operations from which most operations can be derived: omission and insertion. Also for timed automata, mutation operators can be found in Ref. [60].

Mutation analysis for software product lines has not received as much attention yet. In Ref. [61] Stephenson et al. propose the use of mutation testing for prioritizing test cases from a test suite in a product line environment.

Henard et al. proposed two mutation operators for feature models based on propositional formulas in Ref. [62]. They employ their mutation system for showing the effectiveness of dissimilar tests, in contrast to similar tests. For calculating dissimilarity, the authors provide a distance metric to evaluate the degree of similarity between two given products.

7.5 Sampling of products

Sampling product configurations for testing is an ongoing challenge. Most work is focused on structural coverage criteria for feature models and hence is agnostic to the interactions in behavioral models [19, 63]. Still, the test effort is high, since feature interactions are selected for testing where no behavioral interaction is present. Baller et al. focus on heuristics for minimization of the test suite for the base model [64].

Similar to the notion of incremental test design, Beohar et al. propose spinal test suites [65]. A spinal test suite allows one to test the common features of a PL once and for all, and subsequently, only focus on the specific features when moving from one product configuration to another.

Lochau et al. present incremental test design methods to subsequently test every specified behavior [18]. Here, configurations are sampled as needed to achieve the next test goal. The result is a set of test cases where each one is limited to a single product configuration. Our optimization criteria for sampling configurations allow for more flexibility since they use generic test cases.

7.6 Test case assignment

Using three-valued logics for test assessment is an old idea. In 1983, Butler [66] uses the third value to denote either a missing or an incorrect test result. Zhao et al. [67] propose a symbolic three-valued logic analysis to improve the precision of static defect detection. The standardized testing language TTCN-3 [68] allows several verdicts for a test case such as *pass*, *fail*, *inconc*, *none*, and *error*.

To our knowledge, we were the first to propose a theory for a three-valued evaluation of test cases with respect to formal specifications [69]. In Ref. [41], we extended this theory to software product lines based on UML models.

Bertillon et al. [70] use a notation based on natural language descriptions of requirements to define test cases for product lines. The resulting test specification is generic in the product, and a set of relevant test scenarios for a customer specific application can be derived from it. This work complements our coloring method, since we assume that the test suite is designed separately.

8. FUTURE DEVELOPMENTS

Predicting the future of some field always is a risky enterprise, since trends in computer science tend to be short-lived. There are, however, a few clear development tendencies in industry and science which are sketched in this section.

First, we expect product line development methods to become more "main stream" in the engineering of industrial software. Indications of this are, e.g., the availability of tools like pure::variants [71], which interacts with popular development environments such as Mathworks Simulink, IBM Rational DOORS and Rhapsody, and Sparx Systems Enterprise Architect. As another example, FeatureIDE [72] is an open-source tool which tries to "bring all phases of the development process together, consistently and in a user-friendly manner." It supports multiple paradigms for modeling variability such as annotation-based, delta-oriented, and aspect-oriented programming.

Second, there is a trend to standardization. A fist initiative was CVL, the common variability language mentioned above [13]. Since this endeavor has been discontinued, other ways of standardization are being explored. The Variability Exchange Language [73] provides a generic data exchange format for variant management. It is intended to be standardized within OASIS in the near future. The Variability Exchange Language defines a generic API (VariabilityAPI) that allows variant management tools to communicate with artifacts such as model-based specifications, program code, or requirements documents. Since the communication is via a standardized interface, the number of tool adapters is significantly reduced.

Third, the availability of concepts, standards, and unified APIs enables tool providers to focus on the development of features to facilitate the use of their tools. To minimize switching times between different tools and accelerate the overall workflow, a general trend is to integrate tools and support a growing number of other development environments. In particular, for product line development the link to software testing is being strengthened. In this chapter, we discussed the selection of products for testing, and the selection of test cases for products. Integrated tool suites would support finding relevant artifacts for reuse and identifying linked artifacts from other projects. A related research question is how to automatically extract reusable information from development artifacts.

Continuing on the scientific side, of current scientific interest is the refactoring and evolution of software product lines. A question here is how changes in a product line affect subsequent artifacts like implementation models, test cases, and validation and verification documents [74].

Another trend in the research community is to consider models at runtime. With variability models, this would allow for runtime variants and dynamic reconfiguration of systems [75]. For example, one could imagine that additional features, e.g., advanced driver assistants in a car, are loaded according to current demands. Even though these techniques have a high potential, there are numerous safety and security problems to be solved.

Finally, a frequent reconfiguration leads to adaptive and self-learning systems. In fact, dynamic software product lines can be seen as a way to realize adaptive behavior in open contexts. Scientific challenges include the modeling and handling of uncertainties, the integration of feedback in the evolution, and the learning of adaptation rules [76].

9. SUMMARY AND CONCLUSION

In this chapter, we have collected and summarized our recent research on model-based testing for software product lines. With the example application of an eShop product line, we have described how to model a software product line: the feature model describing variation points, the base model describing the functionality, and the variability model which determines which features are realized by which base model elements.

Then, we showed how to automatically derive test suites for product lines from variability models. We distinguished between application-centered and product-centered test generation, and analyzed in detail two approaches of constructing a standard UML model from variability information. A comparison of these approaches showed that both approaches lead to very different test suites. Although they also cover all transitions of the model, domain-centered approaches lead to test suites with significantly fewer test steps. Thus, in terms of test steps and the amount of products to test, domain-centered test design scales better with respect to the size of the system.

We then analyzed the error detection capabilities of generated test suites. For software product lines, there are more sources where errors can occur than in ordinary software models. We defined mutation operators for domain models and measured the detection rate on several examples. The results confirm that operators, which add superfluous behavior to a model, are hard to detect by model-based testing techniques.

Whereas usually in software engineering test suites are built for existing or envisioned software artifacts, in product lines it may be more adequate to build a software artifact for a given test suite. We considered the problem which products should be selected as representatives for a product line, to maximize the benefits from testing within given resource bounds. We found that domain-centered test generation can be very efficient for detecting errors in variability model, and that sampling small and divergent products is better than sampling large ones.

Finally, we looked at a setting where a test suite is developed independently from the product development in a model-based process. This is especially relevant for legacy systems, where a huge set of test cases is already existing, whereas products are restructured along the line. Here, the question is which test cases are applicable for which products. We reduced this problem to a classical model-checking problem and showed how to solve it with existing tools.

While all these contributions can be considered as significant steps, the overall goal of a unified theory for quality assurance of software product lines remains unreached. The main challenge of software product line validation, as compared to "normal" software validation, is the significantly higher complexity. Due to the fact that a product line incorporates a large number of potential products, quality assurance methods must strive to handle many or all of them at once. With our prototypical SPLTestbench, we provide an integrated, open tool environment for this task. In the future, we plan to extend it with capabilities for formal verification and transformational development of software product lines.

REFERENCES

[1] D.L. Parnas, On the design and development of program families. IEEE Trans. Softw. Eng. 2 (1) (1976) 1–9, ISSN 0098-5589, http://dx.doi.org/10.1109/TSE.1976.233797.
[2] J.D. McGregor, L.M. Northrop, S. Jarrad, K. Pohl, Initiating software product lines, IEEE Softw. 19 (4) (2002) 24–27, http://dx.doi.org/10.1109/MS.2002.1020282.
[3] P. Clements, L. Northrop, Software Product Lines: Practices and Patterns, Addison-Wesley, Boston, MA, 2001, ISBN 0-201-70332-7.

[4] K.C. Kang, S.G. Cohen, J.A. Hess, W.E. Novak, A.S. Peterson, Feature-Oriented Domain Analysis (FODA) Feasibility Study 1990, http://www.sei.cmu.edu/library/abstracts/reports/90tr021.cfm.

[5] T. Thüm, C. Kästner, F. Benduhn, J. Meinicke, G. Saake, T. Leich, FeatureIDE: an extensible framework for feature-oriented software development, Sci. Comput. Program. 79 (2014) 70–85, ISSN 0167-6423, http://dx.doi.org/10.1016/j.scico.2012.06.002.

[6] K. Schmid, H. Eichelberger, EASy-producer: from product lines to variability-rich software ecosystems, in: Proceedings of the 19th International Conference on Software Product Line, SPLC '15, Nashville, Tennessee, ACM, New York, NY, 2015, pp. 390–391, http://doi.acm.org/10.1145/2791060.2791112.

[7] S. Apel, D. Batory, C. Kästner, G. Saake, Feature-Oriented Software Product Lines: Concepts and Implementation, Springer-Verlag, Heidelberg, 2013, ISBN 3642375200, 9783642375200.

[8] D. Batory, Feature models, grammars, and propositional formulas, in: Proceedings of the 9th International Conference on Software Product Lines, SPLC'05, Springer-Verlag, Berlin, Heidelberg, ISBN 3-540-28936-4. 2005, pp. 7–20, http://dx.doi.org/10.1007/11554844_3.

[9] K. Czarnecki, M. Antkiewicz, Mapping features to models: a template approach based on superimposed variants, in: R. Glück (Ed.), Generative Programming and Component Engineering, Lecture Notes in Computer Science, vol. 3676, Springer, Berlin [u.a.], ISBN 3-540-29138-5, 2005, pp. 422–437, http://dx.doi.org/10.1007/11561347_28, http://dblp.uni-trier.de/db/conf/gpce/gpce2005.html#CzarneckiA05.

[10] I. Groher, M. Voelter, Expressing feature-based variability in structural models, in: Workshop on Managing Variability for Software Product Lines, 2007.

[11] H. Grönniger, H. Krahn, C. Pinkernell, B. Rumpe, Modeling variants of automotive systems using views, in: T. Kühne, W. Reisig, F. Steimann (Eds.), Tagungsband zur Modellierung 2008 (Berlin-Adlershof, Deutschland, 12–14. März 2008), LNI, Gesellschaft für Informatik, Bonn, 2008.

[12] I. Schaefer, Variability modelling for model-driven development of software product lines, in: D. Benavides, D. Batory, P. Grünbacher (Eds.), ICB-Research Report, Proceedings of the Fourth International Workshop on Variability Modelling of Software-Intensive Systems, Linz, Austria, January 27–29, 2010, ICB-Research Report, vol. 37, Universität Duisburg-Essen, 2010, pp. 85–92.

[13] Ø. Haugen, B. Møller-Pedersen, J. Oldevik, G.K. Olsen, A. Svendsen, Adding standardized variability to domain specific languages, in: SPLC 2008, IEEE Computer Society, Los Alamitos, CA, ISBN 978-0-7695-3303-2, 2008, pp. 139–148, http://dx.doi.org/10.1109/SPLC.2008.25.

[14] A. Memon, Advances in Computers, vol. 86, Academic Press, Cambridge, MA, 2012.

[15] B. Hasling, H. Goetz, K. Beetz, Model based testing of system requirements using UML use case models, in: 1st International Conference on Software Testing, Verification and Validation, 2008, IEEE, Piscataway, NJ, ISBN 978-0-7695-3127-4, 2008, pp. 367–376, http://dx.doi.org/10.1109/ICST.2008.9.

[16] M. Grochtmann, K. Grimm, Classification trees for partition testing, Soft. Test. Verif. Reliab. 3 (2) (1993) 63–82, ISSN 1099-1689, http://dx.doi.org/10.1002/stvr.4370030203.

[17] H. Lackner, Domain-Centered Product Line Testing (Ph.D. thesis), Humboldt-Universität zu Berlin, 2017.

[18] M. Lochau, I. Schaefer, J. Kamischke, S. Lity, Incremental model-based testing of delta-oriented software product lines, in: D. Hutchison, T. Kanade, J. Kittler, J.M. Kleinberg, F. Mattern, J.C. Mitchell, M. Naor, O. Nierstrasz, C. Pandu Rangan, B. Steffen, M. Sudan, D. Terzopoulos, D. Tygar, M.Y. Vardi, G. Weikum, A.D. Brucker,

J. Julliand (Eds.), Tests and Proofs, Lecture Notes in Computer Science, vol. 7305, Springer Berlin Heidelberg, Berlin, Heidelberg, ISBN 978-3-642-30472-9, 2012, pp. 67–82, http://dx.doi.org/10.1007/978-3-642-30473-6_7.

[19] G. Perrouin, S. Sen, J. Klein, B. Baudry, Y. Le Traon, Automated and scalable t-wise test case generation strategies for software product lines, in: ICST '10: International Conference on Software Testing, Verification and Validation, IEEE Computer Society and IEEE, Los Alamitos, CA and Piscataway, N.J, ISBN 978-1-4244-6435-7, 2010, pp. 459–468, http://dx.doi.org/10.1109/ICST.2010.43, http://ieeexplore.ieee.org/stamp/stamp.jsp?arnumber=5477055.

[20] M. Voelter, DSL Engineering: Designing, Implementing and Using Domain-Specific Languages, CreateSpace Independent Publishing Platform, 2013, ISBN 1481218581.

[21] M. Broy, B. Jonsson, J.-P. Katoen, M. Leucker, A. Pretschner (Eds.), Model-Based Testing of Reactive Systems: Advanced Lectures, Lecture Notes in Computer Science, vol. 3472, Springer, Berlin, 2005, ISBN 978-3-540-26278-7, http://dx.doi.org/10.1007/b137241.

[22] H. Götz, M. Nickolaus, T. Roßner, K. Salomon, Modellbasiertes Testen: Modellierung und Generierung von Tests – Grundlagen, Kriterien für Werkzeugeinsatz, Werkzeuge in der Übersicht, iX Studie, vol. 01/2009, Heise Verlag, Hannover, Germany, 2009.

[23] M. Utting, B. Legeard, Practical Model-Based Testing: A Tools Approach, first ed., Morgan Kaufmann Publishers Inc., San Francisco, CA, 2006, ISBN 0123725011.

[24] J. Peleska, Industrial-strength model-based testing—state of the art and current challenges, Electron. Proc. Theor. Comput. Sci. 111 (2013) 3–28, ISSN 2075-2180, http://dx.doi.org/10.4204/EPTCS.111.1.

[25] G. Perrouin, S. Oster, S. Sen, J. Klein, B. Baudry, Y. Traon, Pairwise testing for software product lines: comparison of two approaches, Softw. Qual. J. 20 (3–4) (2012) 605–643, ISSN 0963-9314, http://dx.doi.org/10.1007/s11219-011-9160-9.

[26] S. Weißleder, D. Sokenou, ParTeG—a model-based testing tool, Softwaretechnik-Trends 30 (2) (2010).

[27] M. Genero, D. Miranda, M. Piattini, Defining metrics for UML statechart diagrams in a methodological way, in: G. Goos, J. Hartmanis, J. Leeuwen, M.A. Jeusfeld, Ó. Pastor (Eds.), Conceptual Modeling for Novel Application Domains, Lecture Notes in Computer Science, vol. 2814, Springer, Berlin, Heidelberg, ISBN 978-3-540-20257-8, 2003, pp. 118–128, http://dx.doi.org/10.1007/978-3-540-39597-3_12.

[28] J.A. Cruz-Lemus, A. Maes, M. Genero, G. Poels, M. Piattini, The impact of structural complexity on the understandability of UML statechart diagrams, Inf. Sci. 180 (11) (2010) 2209–2220, ISSN 00200255, http://dx.doi.org/10.1016/j.ins.2010.01.026.

[29] L. Guo, A.S. Vincentelli, A. Pinto, A complexity metric for concurrent finite state machine based embedded software, in: 2013 8th IEEE International Symposium on Industrial Embedded Systems (SIES), 2013, pp. 189–195.

[30] H. Lackner, B.-H. Schlingloff, Modeling for automated test generation—a comparison, in: H. Giese, M. Huhn, J. Phillips, B. Schätz (Eds.), Dagstuhl-Workshop MBEES: Modellbasierte Entwicklung eingebetteter Systeme VIII, Schloss Dagstuhl, Germany, 2012, Tagungsband Modellbasierte Entwicklung eingebetteter Systeme, fortiss GmbH, München, 2012, pp. 57–70.

[31] H. Muccini, A. van der Hoek, Towards testing product line architectures, Electron. Notes Theor. Comput. Sci. 82 (6) (2003) 99–109, ISSN 15710661, http://dx.doi.org/10.1016/S1571-0661(04)81029-6.

[32] M.B. Cohen, M.B. Dwyer, J. Shi, Coverage and adequacy in software product line testing, in: ROSATEA 2006, Association for Computing Machinery, Inc., New York, NY, ISBN 1-59593-459-6, 2006, pp. 53–63, http://dx.doi.org/10.1145/1147249.1147257, http://portal.acm.org/citation.cfm?id=1147249.1147257&coll=Portal&dl=GUIDE&CFID=65839091&CFTOKEN=21681387.

[33] S. Kim, J.A. Clark, J.A. Mcdermid, Class mutation: mutation testing for object-oriented programs, in: FMES, 2000, pp. 9–12.

[34] S.C.P.F. Fabbri, M.E. Delamaro, J.C. Maldonado, P.C. Masiero, Mutation analysis testing for finite state machines, in: 1994 IEEE International Symposium on Software Reliability Engineering, 1994, pp. 220–229.

[35] J. Offutt, S. Liu, A. Abdurazik, P. Ammann, Generating test data from state-based specifications, Softw. Test. Verif. Reliab. 13 (2003) 25–53.

[36] F. Belli, C.J. Budnik, L. White, Event-based modelling, analysis and testing of user interactions: approach and case study, Softw. Test. Verif. Reliab. 16 (1) (2006) 3–32, ISSN 0960-0833, http://dx.doi.org/10.1002/stvr.335.

[37] F. Belli, A. Hollmann, Test generation and minimization with basic statecharts, in: E.J. Delp, P.W. Wong (Eds.), The 2008 ACM Symposium, vol. 5681, SPIE and IS&T, Bellingham, WA and Springfield, VA, ISBN 978-1-59593-753-7, 2008, pp. 718, http://dx.doi.org/10.1145/1363686.1363856.

[38] S. Weißleder, Influencing factors in model-based testing with UML state machines: report on an industrial cooperation, in: D. Hutchison, T. Kanade, J. Kittler, J.M. Kleinberg, F. Mattern, J.C. Mitchell, M. Naor, O. Nierstrasz, C. Pandu Rangan, B. Steffen, M. Sudan, D. Terzopoulos, D. Tygar, M.Y. Vardi, G. Weikum, A. Schürr, B. Selic (Eds.), Model Driven Engineering Languages and Systems, Lecture Notes in Computer Science, vol. 5795, Springer, Berlin, Heidelberg, ISBN 978-3-642-04424-3, 2009, pp. 211–225, http://dx.doi.org/10.1007/978-3-642-04425-0_16.

[39] S. Weißleder, F. Wartenberg, H. Lackner, Automated test design for boundaries of product line variants, in: K. El-Fakih, G. Barlas, N. Yevtushenko (Eds.), Testing Software and Systems, vol. 9447, Springer International Publishing, Cham, ISBN 978-3-319-25944-4, 2015, pp. 86–101, http://dx.doi.org/10.1007/978-3-319-25945-1_6.

[40] K. Kuchcinski, R. Szymanek, JaCoP - Java Constraint Programming Solver, Tech. rep., Lund University, 2013, http://lup.lub.lu.se/record/4092008/file/4092009.pdf.

[41] A. Knapp, M. Roggenbach, B.-H. Schlingloff, On the use of test cases in model-based software product line development, in: Proc. SPLC 2014—18th International Software Product Line Conference. Florence, 2014.

[42] A. Knapp, M. Roggenbach, B.-H. Schlingloff, Automating test case selection in model-based software product line development, Int. J. Softw. Inform. 9 (2) (2015) 153–175 (Special Issue).

[43] Object Management Group, Unified Modeling Language Superstructure. Version 2.4.1, Specification, OMG, 2011, http://www.omg.org/spec/UML/.

[44] A. Knapp, J. Wuttke, Model checking of UML 2.0 interactions, in: Proc. MoDELS 2006 Wsh.s, LNCS 4364, Springer, 2007, pp. 42–51.

[45] G.J. Holzmann, The SPIN Model Checker, Addison-Wesley, Boston, MA, 2003.

[46] R. de Nicola, M. Hennessy, Testing equivalences for processes, Theor. Comput. Sci. 34 (1984) 83–133.

[47] K. Pohl, G. Böckle, Linden, Frank J. van der, Software Product Line Engineering: Foundations, Principles and Techniques, Springer-Verlag New York, Inc., Secaucus, NJ, 2005, ISBN 3540243720.

[48] CVL Revised Submission, http://www.omgwiki.org/variability/doku.php (01.11.15).

[49] P.E. Ammann, J. Offutt, Introduction to Software Testing, Cambridge University Press, New York, NY, 2008, ISBN 9780521880381.

[50] R.V. Binder, Testing Object-Oriented Systems: Models, Patterns, and Tools, Addison-Wesley Longman Publishing Co., Inc, Boston, MA, 1999, ISBN 0-201-80938-9.

[51] G.J. Myers, C. Sandler, T. Badgett, The Art of Software Testing, third ed., John Wiley & Sons, Hoboken, NJ, 2012, ISBN 1118133153.

[52] J. Zander, I. Schieferdecker, P.J. Mosterman, A taxonomy of model-based testing for embedded systems from multiple industry domains, in: J. Zander, I. Schieferdecker,

P.J. Mosterman (Eds.), Model-Based Testing for Embedded Systems, Computational Analysis, Synthesis, and Design of Dynamic Systems, CRC Press, Boca Raton, ISBN 1439818452, 2011.

[53] E. Engström, P. Runeson, Software product line testing—a systematic mapping study, Inf. Softw. Technol. 53 (1) (2011) 2–13.

[54] P.A. da Mota Silveira Neto, I. do Carmo Machado, J.D. McGregor, E.S. de Almeida, S. R. de Lemos Meira, A systematic mapping study of software product lines testing, Inf. Softw. Technol. 53 (5) (2011) 407–423.

[55] B.P. Lamancha, M. Polo, M. Piattini, Systematic review on software product line testing, Commum. Comput. Inf. Sci. 170 (2013) 58–71.

[56] M.G. Burke, M. Antkiewicz, K. Czarnecki, FeaturePlugin, in: J.M. Vlissides, D.C. Schmidt (Eds.), OOPSLA 2004, ACM SIGPLAN Notices, vol. 39(10), Association for Computing Machinery, New York, NY, 2004, pp. 67–72, ISBN 1-58113-831-8.

[57] E.M. Olimpiew, H. Gomaa, Model-based testing for applications derived from software product lines, SIGSOFT Softw. Eng. Notes 30 (4) (2005) 1–7, ISSN 0163-5948, http:// dx.doi.org/10.1145/1082983.1083279.

[58] J.D. McGregor, Testing a Software Product Line, 2001, http://www.bibsonomy.org/ api/users/ist_spl/posts/aec593ff707ee5e036253d36633a414e.

[59] K. Pohl, A. Metzger, Software product line testing, Commun. ACM 49 (12) (2006) 78–81, ISSN 0001-0782, http://dx.doi.org/10.1145/1183236.1183271.

[60] B.K. Aichernig, F. Lorber, D. Ničković, Time for mutants—model-based mutation testing with timed automata, in: D. Hutchison, T. Kanade, J. Kittler, J.M. Kleinberg, F. Mattern, J.C. Mitchell, M. Naor, O. Nierstrasz, C. Pandu Rangan, B. Steffen, M. Sudan, D. Terzopoulos, D. Tygar, M.Y. Vardi, G. Weikum, M. Veanes, L. Viganò (Eds.), Tests and Proofs, Lecture Notes in Computer Science, vol. 7942, Springer, Berlin, Heidelberg, ISBN 978-3-642-38915-3, 2013, pp. 20–38, http://dx.doi.org/ 10.1007/978-3-642-38916-0_2.

[61] Z. Stephenson, Y. Zhan, J. Clark, J. McDermid, Test data generation for product lines—a mutation testing approach, in: B. Geppert, C. Krueger, J. Li (Eds.), SPLiT '04: Proceedings of the International Workshop on Software Product Line Testing, Boston, MA, 2004, pp. 13–18.

[62] C. Henard, M. Papadakis, G. Perrouin, J. Klein, Y. Le Traon, Assessing software product line testing via model-based mutation: an application to similarity testing, in: ICSTW '13: IEEE 6th International Conference On Software Testing, Verification and Validation Workshops 2013, ISBN 978-1-4799-1324-4, 2013, pp. 188–197.

[63] S. Oster, I. Zorcic, F. Markert, M. Lochau, MoSo-PoLiTe: tool support for pairwise and model-based software product line testing, in: VaMoS '11, 2011, pp. 79–82.

[64] H. Baller, S. Lity, M. Lochau, I. Schaefer, Multi-objective test suite optimization for incremental product family testing, in: Proc. ICST 2014, 2014, pp. 303–312, http:// dx.doi.org/10.1109/ICST.2014.43.

[65] H. Beohar, M.R. Mousavi, Spinal test suites for software product lines, in: B.-H. Schlingloff, A.K. Petrenko (Eds.), Proceedings Ninth Workshop on Model-Based Testing, MBT 2014, Grenoble, France, 6 April 2014, EPTCS, vol. 141, 2014, pp. 44–55, http://dx.doi.org/ 10.4204/EPTCS.141.4.

[66] J.T. Butler, Relations Among System Diagnosis Models With Three-Valued Test Outcomes, IEEE, New York, NY, 1983.

[67] Y. Zhao, Y. Wang, Y. Gong, H. Chen, Q. Xiao, Z. Yang, STVL: improve the precision of static defect detection with symbolic three-valued logic, in: APSEC, IEEE Computer Society, 2011, pp. 179–186.

[68] J. Grabowski, D. Hogrefe, G. Rethy, I. Schieferdecker, A. Wiles, C. Willcock, An introduction to the testing and test control notation (TTCN-3), Comput. Netw. 42 (3) (2003) 375–403.

[69] T. Kahsai, M. Roggenbach, B.-H. Schlingloff, Specification-based testing for refinement, in: SEFM 2007—Proc. 5th IEEE International Conference on Software Engineering and Formal Methods, London, 2007.

[70] A. Bertolino, S. Gnesi, Use case-based testing of Product Lines, in: Proc. ESEC/FSE 2003, ACM, 2003, pp. 355–358, http://dx.doi.org/10.1145/940071.940120.

[71] D. Beuche, pure::variants, in: R. Capilla, J. Bosch, K.-C. Kang (Eds.), Systems and Software Variability Management, Springer, Berlin, Heidelberg, ISBN 978-3-642-36582-9, 2013, pp. 173–182, http://dx.doi.org/10.1007/978-3-642-36583-6.

[72] T. Thüm, C. Kästner, F. Benduhn, J. Meinicke, G. Saake, T. Leich, FeatureIDE: an extensible framework for feature-oriented software development, Sci. Comput. Program. 79 (2014) 70–85, ISSN 0167-6423, http://dx.doi.org/10.1016/j.scico.2012.06.002.

[73] M. Gro AŸe Rhode, M. Himsolt, M. Schulze, The Variability Exchange Language, Version 1.0, 2015, Fünfter Workshop zur Zukunft der Entwicklung softwareintensiver, eingebetteter Systeme (ENVISION2020) im Rahmen der SE2015, Dresden, http://www.variability-exchange-language.org/.

[74] V. Alves, R. Gheyi, T. Massoni, U. Kulesza, P. Borba, C. Lucena, Refactoring product lines, in: Proceedings of the 5th International Conference on Generative Programming and Component Engineering, Portland, Oregon, USA, GPCE '06, ACM, New York, NY, ISBN 1-59593-237-2, 2006, pp. 201–210, http://dx.doi.org/10.1145/1173706.1173737.

[75] M. Lochau, J. Bürdek, S. Hölzle, A. Schürr, Specification and automated validation of staged reconfiguration processes for dynamic software product lines. Softw. Syst. Model. 16 (1) (2017) 125–152, ISSN 1619-1374, http://dx.doi.org/10.1007/s10270-015-0470-4.

[76] A.M. Sharifloo, A. Metzger, C. Quinton, L. Baresi, K. Pohl, Learning and evolution in dynamic software product lines, in: Proceedings of the 11th International Symposium on Software Engineering for Adaptive and Self-Managing Systems, Austin, Texas, SEAMS '16,ACM, New York, NY, ISBN 978-1-4503-4187-5, 2016, pp. 158–164, http://dx.doi.org/10.1145/2897053.2897058.

ABOUT THE AUTHORS

Hartmut Lackner Hartmut Lackner's motto: "Life is too short for manual testing," indicates his commitment to automated testing. After finishing his studies in computer science at the Humboldt University of Berlin 2009, he worked at Fraunhofer Institute for Open Communication Systems (Fraunhofer FOKUS) as manager of industrial research projects involving model-based test methods. From 2013 to 2016 he was granted a scholarship at the graduate school METRIK at Humboldt University. During his Ph.D., he specialized in model-based testing of multivariant systems. Since 2016 he is the head of MES Test Center, where he drives the development of testing complex software models.

Bernd-Holger Schlingloff is chief scientist at the Fraunhofer institute FOKUS and professor for software technology at the Humboldt University of Berlin. His main interests are specification, verification, and testing of embedded safety-critical software. He obtained his Ph. D. from the Technical University of Munich in 1990 with a thesis on temporal logic of trees; after that, he worked on model checking of real-time systems. His habilitation in 2001 was on partial state space of safety-critical systems. Since 2002, when he joined Fraunhofer, he is managing industrial projects in the automotive, railway, and medical technology domain. His areas of expertise include quality assurance of embedded control software, model-based development and model checking, logical verification of requirements, software product lines, and automated software testing.

CHAPTER FIVE

Advances in Model-Based Testing of Graphical User Interfaces

Fevzi Belli*, Mutlu Beyazıt[†], Christof J. Budnik[‡], Tugkan Tuglular[§]

*University of Paderborn, Paderborn, Germany
[†]Yaşar University, İzmir, Turkey
[‡]Siemens Corporation, Corporate Technology, Princeton, NJ, United States
[§]Izmir Institute of Technology, Urla, Izmir, Turkey

Contents

1. Introduction: User Interfaces and Their Holistic Testing   220
2. Modeling of GUIs of Interactive Systems   223
   2.1 Different Techniques for Modeling GUIs   224
   2.2 Analysis of Models   229
   2.3 Exploiting Model Morphology for Event-Based Testing   234
   2.4 Test Generation From Models   236
   2.5 Issues to Consider   241
3. Testing and Test Optimization Exemplified by GUI-Modeling With ESG   241
   3.1 Test Termination as an Optimizing Problem   242
   3.2 Exploiting the Structural Features of SUT for Further Reduction of Test Effort   245
   3.3 Case Studies and Their Empirical Evaluation for the Practice   247
4. Contract-Based Testing of GUIs   253
   4.1 Contract-Based Testing in General   254
   4.2 ESG-Based Contract Testing of GUIs   255
5. Rationalization and Automation of GUI Testing   262
   5.1 MBT Tools in General   263
   5.2 Test Tools for GUIs   264
   5.3 Test Automation in Theory and Practice   265
   5.4 Test Tool Requirements in Industry   266
   5.5 ESG Test Suite Designer   267
6. Conclusions   271
   6.1 Peer Into the Future   271
   6.2 Summary   273
References   274
About the Authors   279

Abstract

Graphical user interfaces (GUIs) enable comfortable interactions of the computer-based systems with their environment. Large systems usually require complex GUIs, which are commonly fault prone and thus are to be carefully designed, implemented, and tested. As a thorough testing is not feasible, techniques are favored to test relevant features of the system under test that will be specifically modeled. This chapter summarizes, reviews, and exemplifies conventional and novel techniques for model-based GUI testing.

1. INTRODUCTION: USER INTERFACES AND THEIR HOLISTIC TESTING

There are two distinct types of construction work while developing software:

- Design, implementation, and test of the programs.
- Design, implementation, and test of the user interface (UI).

We assume that UI might be constructed separately, as it requires different skills, and maybe different techniques than construction of common software. The design part of the development job requires a good understanding of user requirements; the implementation part requires familiarity with the technical equipment, i.e., programming platform, language, etc. Testing requires both a good understanding of user requirements and familiarity with the technical equipment. This chapter is about UI testing, i.e., testing of the software that realizes the UI, taking the design aspects into account. To some extent, also analysis aspects are covered because testing and analysis usually belong together.

Graphical user interfaces (GUIs) have become more and more popular and common UIs in computer-based systems. Testing GUIs is, on the other hand, a difficult and challenging task for many reasons: First, the input space possesses a great, potentially infinite number of combinations of inputs and events that occur as system outputs; external events may interact with these inputs. Second, even simple GUIs possess an enormous number of states which are also due to interaction with the inputs. Last but not least, many complex dependencies may hold between different states of a GUI system, and/or between its states and inputs.

User inputs are critical for the security, safety, and reliability of software systems. As Whittaker [1] indicated, "... data is the lifeblood of software; when it is corrupt, the software is as good as dead." According to Whittaker, this is indeed the bottom line for software developers and testers. One must consider every single input from every external resource to have confidence in the ability of the system under test (SUT) to properly handle malicious attacks and unanticipated operating environments. Deciding which inputs to trust and which to validate is a constant balancing act. Experiences from safety and security fields [1] have shown that user inputs, mostly obtained from GUIs, should be validated thoroughly to prevent attacks ranging from injection to denial of service and resulting in intrusion or even in system crashes. The same is true for safety violations.

Nevertheless, nowadays it is taken for granted that most human–computer interfaces (HCIs) are materialized via GUIs. There exists a vast amount of research work for specification of HCIs, resulting in an effective testing strategy which is not only easy to apply but also scalable in sense of stepwise increasing the test complexity and accordingly the test coverage and completeness of the test process, thus also stepwise increasing the test costs in accordance with the test budget of the project. There has been, however, little well-known systematic study in this field. This chapter presents techniques to systematically test GUIs, being capable of test case enumeration for precise test scalability. Aspects of test optimization and rationalization by tools are also covered.

Test cases generally require the determination of meaningful test inputs and expected system outputs for these inputs. Accordingly, to generate test cases for a GUI, one has to identify the test objects and test objectives. The test objects are the instruments for the input, e.g., screens, windows, icons, menus, pointers, commands, function keys, alphanumerical keys, etc. The objective of a test is to generate the expected system behavior (desired event) as an output by means of well-defined test input, or inputs. In a broader sense, the test object is the software under test (SUT); the objective of the test is to gain confidence in the SUT. Robust systems possess also a good exception handling mechanism, i.e., they are responsive not in terms of behaving properly in case of correct, legal inputs, but also by behaving good natured in case of illegal inputs, generating constructive warnings, or tentative correction trials, etc., that help to navigate the user to move in the right direction. In order to validate such robust behavior, one needs systematically generated erroneous inputs which would usually entail injection of undesired events into the SUT. Such events would usually transduce the software under test into an illegal state, causing even a system crash, if the program does not possess an appropriate exception handling mechanism. This is called "negative testing" and is also subject of this chapter.

Test inputs of a GUI usually represent sequences of GUI-object activities and/or selections that operate interactively with the objects (Interaction Sequences—IS, [2], see also Refs. [3,4]). Among these ISs, the ones interesting for testing are externally observable (Event Sequences—ES). Such an

ES is complete (CES), if and only if it eventually invokes the desired system responsibility. From the Knowledge Engineering point of the view, the testing of GUI represents a typical planning problem that can be solved goal-driven [4]: Given a set of operators, an initial state and a goal state, the planner is expected to produce a sequence of operators that change the initial state to the goal state. For the GUI test problem described earlier, this means we have to construct the test sequences in dependency of both the desired, correct events (positive testing) and the undesired, faulty events (negative testing). A major problem is the unique distinction between correct and faulty UI events (Oracle Problem [5,6]). The chapter reviews approaches that exploit the concept of ES to elegantly cope with the Oracle Problem.

GUI testing can be performed using a model-based testing (MBT) approach. In MBT, a model describing the behavior of SUT is created and this model is used for automatic generation of test cases which are then applied to the SUT. The basic idea is to use some coverage criteria to generate test cases [4–6]. Achieving a proper level of coverage entails the generation of test cases and the selection of an optimal number of them. Thus, it ensures the cost-effective exercise of a given set of structural or functional features.

Another tough problem while testing is the decision when to stop testing (Test Termination Problem and Testability [5,6]). Exercising a set of test cases, the test results can be satisfactory, but this is limited to these special test cases. Thus, for the quality judgment of the SUT, one needs further, rather quantitative arguments, usually materialized by well-defined coverage criteria. The most well-known coverage criteria are based on either special, structural issues of the software to be tested (implementation orientation/white-box testing), or its behavioral, functional description (specification orientation/black-box testing), or both, if both implementation and specification are available (hybrid/gray-box testing).

Putting the different components of the approach together, a holistic way of modeling of software development is materialized, with the novelty that the complementary view of the desired system behavior enables to obtain the precise and complete description of undesired situations, leading to a systematic, scalable, and complete fault modeling.

This chapter summarizes existing work on model- and specification-based, positive and negative GUI testing, depicting it by many examples, from simple ones, e.g., copy/cut-paste process, to examples lent from public domain Internet, e.g., Real Jukebox (an interactive personal music management program), and to examples lent from real projects, e.g., from the automotive industry, namely a proactive system to control a marginal strip mower mounted on a truck.

This chapter prefers a graph-based modeling technique as it is widely accepted in the practice. Moreover, formal methods, for example, graph theory, can be applied to graph-based models for achieving algorithms for design, validation & verification, and optimization. All sections use, exemplarily, ESG modeling. For enabling quantification of test cases ESG modeling will be augmented by decision tables. Moreover, the idea of "model morphology" will be introduced from mutation analysis and testing to refine the holistic view.

Section 2 introduces the notion of finite-state modeling centered mainly on event-based models which are most commonly used for both modeling the system and the faults through sequence relation between the events. Notions and results from mutation analysis and testing have been adopted for test case generation. Cost aspects are discussed in Section 3 that introduces an optimization model to solve the test termination problem. Some potentials of test cost reduction are discussed. A new approach, GUI testing based on contracts and decision tables, is introduced in Section 4. Section 5 reviews existing tools for GUI testing, before Section 6 concludes the chapter, considering also future research directions.

Techniques represented in this chapter are to a great extend based on sound mathematical methods. Thus, they enable a formal, algorithmic handling, and consequently automatization.

The authors decided to integrate related work into the relevant sections of the chapter instead of having an extra section. This simplifies the work and avoids multiple referencing, that is, once in relevant section(s), and another time in the extra section on related work.

2. MODELING OF GUIs OF INTERACTIVE SYSTEMS

In terms of behavioral patterns, the relationships between the SUT and its environment, i.e., the user, the natural environment, etc., can be described as proactive, reactive, or interactive. In the case of proactivity, the system generates the stimuli, which evoke the activity of its environment. In the case of reactivity, the system behavior evolves through its responses to stimuli generated by its environment. Most HCIs are nowadays interactive in the sense that the user and the system can be both pro- and reactive. In this particular kind of system, the UI can have another environment, that is, the plant, the device, or the equipment, which the UI is intended for and embedded in, controlling technical processes. Modern UIs will be mostly implemented graphically; therefore, we will concentrate on graphical, interactive user interfaces (GUI); UI and GUI will be used interchangeably.

This section discusses the modeling of GUIs for the purpose of testing. In Section 2.1, different GUI modeling techniques are introduced; the discussion is mainly centered around event-based models such as event flow graphs (EFGs), event sequence graphs (ESGs), and k-sequence right regular grammars (k-Regs). In Sections 2.2–2.4, the discussion is based on a generic event-based modeling methodology using k-Regs. In Section 2.2, the basic notions and properties needed for testing purposes and fault-based aspects are analyzed. In Section 2.3, the concept of varying model morphology and its benefits are discussed, and in Section 2.4, test generation perspectives are demonstrated. Section 2 is concluded by Section 2.5 by laying out and discussing some issues related to the MBT methodologies.

2.1 Different Techniques for Modeling GUIs

An event is defined as an externally observable phenomenon, e.g., a user, a stimulus, or a response of the GUI, punctuating different stages of the system activity. Event-based modeling techniques are commonly used for modeling GUIs. The most well-known event-based techniques are centered on the use of ESGs [7], EFGs [8], and a particular type of regular grammars called k-Regs [9,10].

Although we focus on event-based models, state-based or algebraic models can also be used for testing of GUIs. FSMs [4,11,12], variable FSMs [13], hierarchical FSMs [14], statecharts [15,16], and pushdown automata [17,18] are common examples to state-based models, whereas regular expressions [19] can be considered as algebraic.

2.1.1 Event Sequence Graphs

An *ESG* is a 4-tuple (N, A, S, F) where

- N is a finite set of nodes representing the *events*.
- $A \subseteq N \times N$ is a finite set of directed arcs representing *follows* relation between events, that is, for two events x and y in the graph, x *follows* y if and only if (y, x) is an arc in the graph.
- $S \subseteq N$ is a distinguished nonempty set of events representing *start* or *initial* events.
- $F \subseteq N$ is a distinguished nonempty set of events representing *finish* or *final* events.

The above definition suggests that ESG-based modeling employs a very simplistic approach. Each node in an ESG is an event whose type- or application-specific semantics are ignored; and each arc represents a sequence of two events.

An example GUI is given in Fig. 1. It contains a total of 11 events (all taskbar and non-GUI-related events are ignored): "Cut," "Copy," "Paste," "Find...," "Go To...," "Find what," "Find Next," "Cancel (Find)," "Line Number," "OK," and "Cancel (GoTo)." Performing "Go To..." and "Find..." events bring forth subcomponents of the GUI enabling the execution of different events. After "Find...," "Find what," "Find next," and "Cancel (Find)" and, after "Go To...," "Line Number," "OK," and "Cancel (GoTo)" events are enabled.

Fig. 2 is an ESG model of the GUI given in Fig. 1. Events are only distinguished as simple and composite events: Simple events are shown in ellipses and they correspond to actual events. The composite events ("Go To..." and "Find..."), which have their own ESGs, are shown in the dotted ellipses. Note that the actual "Go To..." and "Find..." events are in fact in their

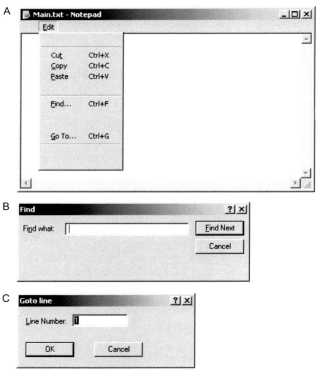

Fig. 1 An example GUI (simplified from Notepad) [20]. (A) "Main" window. (B) "Find" window (modeless). (C) "GoTo line" window (modal).

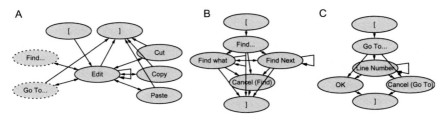

Fig. 2 A simplified ESG model of the GUI in Fig. 1 [20]. (A) "Main" ESG. (B) "Find" sub-ESG. (C) "GoTo line" sub-ESG.

corresponding sub-ESGs. In addition, in all ESGs, pseudoevents "[" and "]" are used to mark start and finish events, respectively; that is, start events follow "[," and "]" follows each finish event.

Note that, although not suggested by the definition, some events in the example ESG are composite; they represent sub-ESGs. This is a quite commonly employed technique in practice to ease the modeling. Such ESGs are actually called *structured* ESGs [21,22] where certain nodes can be refined as long as the refinement is compatible with the notion of event. Similarly, it is possible to extend ESGs to enable the modeling of more complex situations as follows [21,22]:

- *Input–output* ESGs: Input and output events are distinguished semantically.
- *Communicating* ESGs: Two ESGs can communicate with each other to accomplish a task.
- *Quiescent* ESGs: A special event to represent the occurrence of no input or output system action is also included.
- *Timed* ESGs: Event-based behavior is defined with respect to time.
- *Pushdown* ESGs: A stack component is included as a special type of memory.

2.1.2 Event Flow Graphs

An *EFG* [8], on the other hand, takes component-based structure of the GUIs into account and distinguishes between different types of events such as menu-open events, restricted-focus events, unrestricted focus events, termination events, and system-interaction events [23]. In this perspective, EFGs can be considered as an extension to ESGs where node semantics are augmented using application-specific details. An EFG for a GUI component C is a four-tuple (V, E, B, I) where

- V is a set of vertices representing all the *events* in the component. Each $v \in V$ represents an event in C where it can be a menu-open, a restricted-focus, a termination, or a system-interaction event.

- $E \subseteq N \times N$ is a set of *directed edges* between vertices. We say that event e_i follows e_j if and only if e_j may be performed *immediately* after e_i. An edge $(v_x, v_y) \in E$ if and only if the event represented by v_y *follows* the event represented by v_x.
- $B \subseteq V$ is a set of vertices representing those events of C that are available to the user when the component is first invoked.
- $I \subseteq V$ is the set of restricted-focus events of the component.

Fig. 3 shows an example EFG model of the GUI in Fig. 1. The events are distinguished based on their types and different shapes are used for such events in modeling. "Edit" is a menu–open event (shown in diamond), "Go To…" is a restricted-focus event (shown in double circle), and related "OK" and "Cancel (GoTo)" are termination events (shown in rectangles). Also, "Find…" is an unrestricted focus event (because "Find" window is modeless) enabling "Find what" and "Cancel (Find) events, and "Cut," "Copy," and "Paste" events are system–interaction events (shown in circles). Furthermore, "Edit" is designated as the only start event.

EFGs also have different extensions:

- Event Interaction Graphs (EIGs) [24]: An EIG focuses on system interaction and termination events (assuming that other events are not fault prone), and interactions between them.
- Event Semantic Interaction Graphs (ESIGs) [25]: An ESIG models a subset of *follows* relation between events that are shown to interact at a certain semantic level.

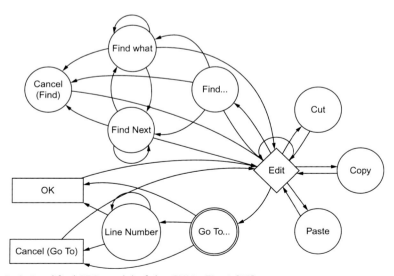

Fig. 3 A simplified EFG model of the GUI in Fig. 1 [20].

- Probabilistic EFGs (PEFGs) [26]: In a PEFG, by analyzing usage profiles, weights or probabilities are assigned to the events to form Bayesian networks and n-gram Markov models.

2.1.3 k-Sequence Right Regular Grammars

In order to model the relations between sequences of events and single events, a generalized family of event-based models is introduced as k-Regs ($k \geq 1$) [9,10,27]. Mainly, k-Regs are defined with a general event-based modeling methodology in mind (although certain elements in the model can also be regarded as states and, thus, enable adoption of state-based approaches). Therefore, event semantics are not specialized using application-specific details. Furthermore, certain ESG-based formalization issues are solved resulting in improvements in the event-based testing approaches. A k-Reg ($k \geq 1$) is a six-tuple (E, B, K, C, S, P) where:

- E is a finite set of *events* (or *contexted events*).
- B is a finite set of *basis events*, which is the set of all visible events under consideration. For each event $e \in E$, $d(e) \in B$ is the corresponding basis event, which is the noncontexted version of e, and $d(.)$ is the *decontexting function*.
- $K \subseteq E^k$ is a finite set of *k-sequences*. For each k-sequence $r \in K$, $r = r_1 \ldots r_k$ and $d(r) = d(r_1) \ldots d(r_k) \in B^k$ is the corresponding *basis k-sequence*.
- C is a finite set of *contexts* where $S \in C$ is the *start context*.
- P is a finite set of *productions* of the form

$$Q \rightarrow \varepsilon \ \text{ or } \ Q \rightarrow r\,c(r)$$

where $Q \in C$ is a context, $r \in K$ is a k-sequence, $c(r) \in C \setminus \{S\}$ is the unique context of r, and ε is the empty string. If $k \geq 2$, for each $c(q) \rightarrow r\,c(r) \in P$, the ending $(k-1)$-sequence of q is the beginning $(k-1)$-sequence of r.

The semantics of the productions of a k-Reg G is as follows:

- For each $c(q) \rightarrow r\,c(r) \in P$, where $q = q_1 \ldots q_k$ and $r = r_1 \ldots r_k$, *r follows q in grammar G*, and r_k *follows q in the system modeled by grammar G*; that is, $q_1 \ldots q_k\,r_k$ is a $(k+1)$-sequence in the system.
- For each $S \rightarrow r\,c(r) \in P$, r is a *start k-sequence*.
- For each $c(q) \rightarrow \varepsilon \in P$, q is a *finish k-sequence*.

Fig. 4 shows an example 1-Reg model of the GUI in Fig. 1. Productions of the form $S \rightarrow r\,c(r)$ and $c(r) \rightarrow \varepsilon$ are used to mark start events and finish events, respectively. Furthermore, productions of the form $c(q) \rightarrow r\,c(r)$ are used to model the *follows* relation between single events. Therefore, it is possible to represent the productions visually using directed graphs (similar to ESGs).

```
 1. S → Edit c(Edit)                              20. c(LineNumber) → Cancel(GoTo) c(Cancel(GoTo))
 2. c(Edit) → ε                                   21. c(OK) → ε
 3. c(Edit) → Edit c(Edit)                        22. c(OK) → Edit c(Edit)
 4. c(Edit) → Cut c(Cut)                          23. c(Cancel(GoTo)) → ε
 5. c(Edit) → Copy c(Copy)                        24. c(Cancel(GoTo)) → Edit c(Edit)
 6. c(Edit) → Paste c(Paste)                      25. c(Find...) → Edit c(Edit)
 7. c(Edit) → GoTo... c(GoTo...)                  26. c(Find...) → Findwhat c(Findwhat)
 8. c(Edit) → Find... c(Find...)                  27. c(Find...) → FindNext c(FindNext)
 9. c(Cut) → ε                                    28. c(Find...) → Cancel(Find) c(Cancel(Find))
10. c(Cut) → Edit c(Edit)                         29. c(Findwhat) → Edit c(Edit)
11. c(Copy) → ε                                   30. c(Findwhat) → Findwhat c(Findwhat)
12. c(Copy) → Edit c(Edit)                        31. c(Findwhat) → FindNext c(FindNext)
13. c(Paste) → ε                                  32. c(Findwhat) → Cancel(Find) c(Cancel(Find))
14. c(Paste) → Edit c(Edit)                       33. c(FindNext) → Edit c(Edit)
15. c(GoTo...) → LineNumber c(LineNumber)         34. c(FindNext) → Findwhat c(Findwhat)
16. c(GoTo...) → OK c(OK)                         35. c(FindNext) → FindNext c(FindNext)
17. c(GoTo...) → Cancel(GoTo) c(Cancel(GoTo))     36. c(FindNext) → Cancel(Find) c(Cancel(Find))
18. c(LineNumber) → LineNumber c(LineNumber)      37. c(Cancel(Find)) → Edit c(Edit)
19. c(LineNumber) → OK c(OK)
```

Fig. 4 A simplified 1-Reg model of the GUI in Fig. 1.

ESGs, EFGs, and k-Regs are all used to define a *follows* relation by properly identifying start and finish events. In case of ESGs and EFGs, this relation is between events; however, in case of k-Regs, this relation is between k-sequences and events. From a formal point of view, ESGs, EFGs, and 1-Regs are similar to FSMs. By loosening the constraints on start and finish events, one can convert an ESG, an EFG, or a 1-Reg to an FSM by interpreting the formers as Moore-like machines [28] and the latter as Mealy-like machine [29], and vice versa. However, the case of empty string and the absence of final states should be handled carefully. Furthermore, use of indexing (or contexting) [7] may be necessary to assign a unique label to each event.

2.2 Analysis of Models

In general, when a model is given, different types of analyses can be carried out. Considering the aforementioned event-based models, for event-based testing of GUIs, it is possible to perform an analysis to test for *legal* (*valid*, *desirable*, *correct*, or *positive*) and *illegal* (*invalid*, *undesirable*, *faulty*, or *negative*) *system behaviors*, that is, the behaviors allowed and not allowed by the system, respectively. In this section, in order to represent such behaviors and carry out our analysis, the discussion is based on certain notions borrowed from mutation analysis and testing [30–32].

2.2.1 Basic Notions

Let G be a k-Reg. Event sequences that can and cannot be derived using G are distinguished for testing. An event sequence s is said to be *in* G, if it can be derived using the productions (or the *follows* relation) in G. A nonempty event sequence s in G is a *start sequence*, if it starts with a start event; s in G is a *finish sequence*, if it ends with a finish event.

A

S → c1 c(c1) | x1 c(x1)
c(c1) → c1 c(c1) | x1 c(x1) | p1 c(p1)
c(x1) → c1 c(c1) | x1 c(x1) | p2 c(p2)
c(p1) → c1 c(c1) | x1 c(x1) | p1 c(p1) | ε
c(p2) → c1 c(c1) | x1 c(x1) | ε

B

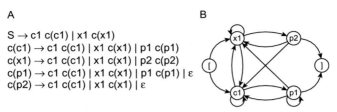

Fig. 5 A 1-Reg model [10]. (A) Productions. (B) Directed graph visualization.

Fig. 5 shows an example 1-Reg where *c* is *copy*, *x* is *cut*, and *p* is *paste* basis event, and $c1, x1, p1$, and $p2$ are contexted versions of these events. $\{c1\ x1, c1\ p1\ p1, p2\ x1\ p2, p1\ p1\ p1\ c1\}$ is an example set which contains some sequences in this 1-Reg, whereas $\{c1\ p2, x1\ p1\ p2, p2\ p2\ x1\ c1\}$ contains some sequences not in it. Note that, in the figure, productions of the form $H \rightarrow T1, H \rightarrow T2, ..., H \rightarrow TN$ are grouped as $H \rightarrow T1 \mid T2 \mid ... \mid TN$ to save space.

An event sequence in G can be used to exercise some desirable or correct behavior, whereas an event sequence not in G can be used to exercise some undesirable or faulty behavior and it is also called as a *faulty event sequence* (*FES*). Original model G and its mutants can be used to generate positive and negative test cases for these purposes. More precisely, the aim is to reveal *missing event faults* where an event cannot occur after or before a (possibly empty) sequence of events and *extra event faults* where an event can occur after or before a (possibly empty) sequence of events.

An event sequence is a *positive test case*, if it is a start sequence in G (or it is ε). $T_P(G)$ denotes the *set of all positive test cases*. A *complete event sequence* (*CES*) is a positive test case which is both a start and a finish sequence in G (or it is ε). $T_{CES}(G) \subseteq T_P(G)$ denotes the *set of all CESs*. Furthermore, an event sequence is a *negative test case*, if the first event in it is a nonstart event or it contains at least one 2-sequence which is not in M. $T_N(G)$ denotes the *set of all negative test cases*. A *faulty complete event sequence (FCES)* is a negative test case which either is composed of only a nonstart event or contains only one 2-sequence which is not in G and it ends with this 2-sequence. $T_{FCES}(G) \subseteq T_N(G)$ denotes the *set of all FCESs*.

2.2.2 Relevant Mutants

In general, one can create infinitely many mutants modeling multiple missing event or extra event faults. For this purpose, *marking (mark start, mark finish, mark nonstart,* and *mark nonfinish), insertion (insert sequence, insert k-sequence),* and *omission (omit sequence* and *omit k-sequence)* operators can be defined [33] by extending the operators defined in Ref. [34]. In the light of the following assumptions which are generally valid for GUI testing, the types and the numbers of mutants can be reduced greatly [9,10]:

A1. Events in a test case are executed in the given order; therefore, execution of a test case stops when a failure is observed.

A2. The last event of a test sequence can be any event; a test sequence needs not to end with a finish event.

Thus, for a given k-Reg, the following can be stated:

P1. Missing and extra event faults are limited by considering the k-sequences which precede the missing or extra events while ignoring the succeeding k-sequences. Thus, by exercising all $(k+1)$-sequences in the k-Reg, one can test whether an event is missing after some k-sequence, and, by exercising all relevant faulty k-sequences, one can test whether an event is extra after some k-sequence (by A1).

P2. Mark nonstart, mark nonfinish, omit sequence, and omit k-sequence mutants are discarded; they do not contain any k-sequence that is not contained in the original model (by P1).

P3. Mark finish and mark nonfinish mutants do not really correspond to fault models because every event can be considered as a finish or a non-finish event during the testing process (by A2).

P4. Insert sequence mutants are discarded because extra event faults modeled using insert sequence mutants can be modeled using insert k-sequence mutants (by definition [33]).

P5. There is no need to continue execution of a negative test case beyond the first faulty sequence. Thus, all negative test cases used in testing process are FCES (by A1).

Consequently, we do not need to use all types of mutants for test generation; we can use the original k-Reg to cover $(k+1)$-sequences to reveal missing event faults, and mark start and insert k-sequence mutants to cover faulty $(k+1)$-sequences to reveal extra event faults. Basically, the purpose of using a mark start operator is to turn a given k-sequence into a start k-sequence. In this way, mutant models that have extra start k-sequences can be constructed. These models are used to generate test cases to reveal extra event faults where the extra event is a start event in a start k-sequence. The operator is defined as follows.

Given a k-Reg $G=(E, B, K, C, S, P)$ and a k-sequence $e \in K$ such that $S \rightarrow e\ c(e) \notin P$, *mark start* (*Ms*) operator is defined as

$$Ms(G, e) = (E, B, K, C, S, P')$$

where $P' = P \cup \{S \rightarrow e\ c(e)\}$.

Insert k-sequence operator adds a new k-sequence to a given grammar following an existing k-sequence. In this way, mutant models which contain k-sequences with different contexts can be created. Such models are used to generate test cases to reveal extra event faults where the extra event follows a k-sequence. The operator is defined as follows.

Given a k-Reg $G = (E, B, K, C, S, P)$, a k-sequence e such that $e \notin K$ and $d(e) \in B^k$, and an existing k-sequence $a \in K$, *insert k-sequence (It)* operator is defined as

$$It(G, e, a) = G' = (E, B, K', C', S, P'),$$

where $K' = K \cup \{e\}$, $C' = C \cup \{c(e)\}$, and $P' = P \cup \{c(a) \to e\, c(e),\ c(e) \to \varepsilon\}$.

Fig. 6 shows a mark start and an insert 1-sequence mutant of the 1-Reg in Fig. 5. (To save space, only the directed graph visualizations are included.)

2.2.3 Mutant Selection

In event-based testing (under assumptions A1 and A2), not all mutants need to be generated. Therefore, having defined the relevant mutation operators, two strategies are defined for mark start and insert k-sequence mutants to select a subset of all possible mutants in such a way that

- each selected mutant models a small number of faults which are located at the mutation points so that one modeled fault does not interfere with another,
- there is no need to compare each mutant to the original model to check for equivalence or to generate test cases to reveal the faults,
- the generation of equivalent mutants and multiple mutants modeling the same faults are avoided, and
- a test case to reveal the fault modeled by the mutant can be generated in linear time.

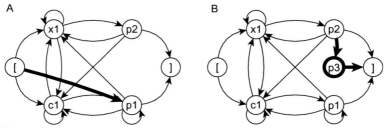

Fig. 6 A mark start and an insert 1-sequence mutant of the 1-Reg in Fig. 5 [10]. (A) $p1$ is marked as start. (B) $p3$ (a p event) is inserted after $p2$.

Given a k-Reg $G=(E, B, K, C, S, P)$, mark start and insert k-sequence mutant selection strategies are defined as follows.

Mark Start Mutant Selection Strategy: For each mark start mutant $Ms(G, e)$ of G, k-sequence e is selected as a mutation parameter if the following conditions hold:

1. There exists no start k-sequence x such that $d(x_1)=d(e_1)$.
2. There exists no previously selected mutation parameter y such that $d(y_1)=d(e_1)$.

Insert k-sequence Mutant Selection Strategy: For each insert k-sequence mutant $It(G, e, a)$ of G, k-sequence e and k-sequence a are selected as a mutation parameter if the following conditions hold:

1. There exists no $c(a) \to x \ c(x) \in P$ such that $d(x_k)=d(e_k)$.
2. There exists no previously selected mutation parameter (y, a) such that $d(y_k)=d(e_k)$.

The algorithms to generate all mark start and insert k-sequence mutants according to the above strategies are given in Algorithms 1 and 2, respectively.

Algorithm 1. Mark Start Mutant Selection

```
Input: G = (E, B, K, C, S, P) - the input grammar
Output: M - the set of selected mark start mutants
M =∅, N = ∅
 for each b ∈ B do
    if there is no S → x c(x) ∈ P such that d(x₁) = b and
        there is no y∈ N such that d(y₁) = b then
         Select a k-sequence e ∈ K such that d(e₁) = b
         G' = G
         M = M ∪ {Ms(G', e)}
         N = N∪ {e}
     endif
 endfor
```

Algorithm 2. Insert k-Sequence Mutant Selection

```
Input: G = (E, B, K, C, S, P) - the input grammar
Output: M - the set of selected insert k-sequence mutants
M =∅, N = ∅
 for each a ∈ K do
     N = ∅
```

```
for each b ∈ B do
    if there is no c(a) → x c(x) ∈ P such that d(xₖ) = b and
        there is no (a, y) ∈ N such that d(yₖ) = b then
            b' = a new contexted version of b, e = a₂ ... aₖ b'
            G' = G
            M = M ∪ {It(G', e, a)},
            N = N ∪ {(a, e)}
    endif
endfor
endfor
```

Let G be the 1-Reg in Fig. 5. The only selected mark start mutant is $Ms(G, p1)$.
$Ms(G, c1)$ and $Ms(G, x1)$ are excluded because $c1$ and $x1$ are already start
events. Furthermore, $Ms(G, p2)$ is excluded because it models the same fault
as $Ms(G, p1)$. Furthermore, one can only use basis 1-sequence p to generate
insert 1-sequence mutant, because c and x can follow all events. The only
selected insert 1-sequence mutant is $It(G, p3, p2)$, because only $p2$ is not
followed by a p event.

2.3 Exploiting Model Morphology for Event-Based Testing

Varying model morphology can be formalized using k-Regs [9,10]. However, not to bore the reader with too much formalism, we keep the discussion semiformal and skip certain details. The interested reader may refer to
Refs. [10,33] for a more complete discussion.

A $(k + 1)$-Reg model is morphologically different from a k-Reg model,
and it can be used to model or reveal different or more subtle faults. For this
purpose, a transformation to vary k and generate models with morphological
differences is defined as follows.

Given a 1-Reg $G_1 = (E, B, K_1, C_1, S, P_1)$.

- *The corresponding 1-Reg of G_1 is defined as itself:*

$$G_1 = (E, B, K_1, C_1, S, P_1).$$

- Let $G_k = (E, B, K_k, C_k, S, P_k)$ be *the corresponding k-Reg of G_1. The
corresponding $(k + 1)$-Reg of G_1 (or G_k) is defined as*

$$G_{k+1} = (E, B, K_{k+1}, C_{k+1}, S, P_{k+1}), \quad \text{where}$$

○ $K_{k+1} = \{q_1 \ldots q_k \, r_k \mid c(q) \rightarrow r \, c(r) \in P_k,$ where $q = q_1 \ldots q_k$ and $r = r_1 \ldots r_k\}$ is the set of all $(k+1)$-sequences in G_1.

○ $C_{k+1} = \{c(r) \mid r \in K_{k+1}\}$ is the set of contexts.

○ $P_{k+1} = \{S \rightarrow r \, e \, c(r \, e) \mid S \rightarrow r \, c(r) \in P_k$ and $c(r_k) \rightarrow e \, c(e) \in P_1\} \cup \{c\text{-}(qr_k) \rightarrow \varepsilon \mid c(q) \rightarrow r \, c(r) \in P_k$ and $c(r_k) \rightarrow \varepsilon \in P_1\} \cup \{c(q \, r_k) \rightarrow r \, e \, c(r \, e) \mid c(q) \rightarrow r \, c(r) \in P_k$ and $c(r_k) \rightarrow e \, c(e) \in P_1\}$ is the set of productions.

The above definition is recursive; it can easily be converted to an iterative algorithm whose steps are outlined in Algorithm 3.

Algorithm 3. k-Reg Transformation

```
Input: Gₖ = (E, B, Kₖ, Cₖ, S, Pₖ) - the input k-Reg (the corresponding
       k-Reg of G₁)
       G₁ = (E, B, K₁, C₁, S, P₁) - the input 1-Reg
Output: Gₖ₊₁ = (E, B, Kₖ₊₁, Cₖ₊₁, S, P ₖ₊₁) - the corresponding
        (k+1)-Reg
   Kₖ₊₁ =∅, Cₖ₊₁ = {S}, Pₖ₊₁ = ∅
   for each Q → r c(r) ∈ Pₖ where r = r₁ ... rₖ do
       if Q = c(q) where q = q₁ ... qₖ then
           Kₖ₊₁ = Kₖ₊₁ ∪ {q rₖ}
           Cₖ₊₁ = Cₖ₊₁ ∪ {c(q rₖ)}
       endif
       for each c(rₖ) → R ∈ P₁ do
           if R = e c(e) then
               if Q = S then
                   Pₖ₊₁ = P ₖ₊₁ ∪ {S → r e c(r e)}
               else if Q = c(q) then
                   Pₖ₊₁ = P ₖ₊₁ ∪ {c(q rₖ) → r e c(r e)}
               endif
           else if R = ε then
               Pₖ₊₁ = Pₖ₊₁ ∪ {c(q rₖ) → ε}
           endif
       endfor
   endfor
```

Based on the definition, Algorithm 3 uses a 1–Reg and a k–Reg to obtain a $(k+1)$–Reg. Basically, what happens is as follows:

• A new $(k+1)$-sequence $q_1 \ldots q_k \, r_k$ in G_1 is extracted from $c(q_1 \ldots q_k) \rightarrow r_1 \ldots r_k \, c(r_1 \ldots r_k) \in P_k$ using the fact that $q_1 \ldots q_k$ is a k-sequence in G_1 and $q_k \, r_k$ is a 2-sequence in G_1.

- To determine the contexts to be used in a new production properly, a production from G_k and a production from G_1 are selected and used in such a way that $(k+1)$-sequences that are not in G_1 does not emerge, and all $(k+1)$-sequences in G_1 are included in new productions together with their contexts without invalidating the definition of a k-Reg.
- k-Sequences in G_1 which cannot be included in some $(k+1)$-sequences in G_1 are left out.

Fig. 7 is the corresponding 2-Reg transformed from the 1-Reg in Fig. 5.

One of the most important benefits of using of morphologically different models, generated using grammar transformation, is the extension of the set of possible mutants (or fault models). To see this, consider the mutant of Fig. 7 generated by omitting sequence ($p1$ $c1$, $c1$ $p1$) as shown in Fig. 8B. This mutant models the fault that

$$p1 \text{ is missing after } p1\,c1,$$

that is, p fails after performing a p and a c. It is not possible to create such a mutant from the model in Fig. 5 by a simple omission. For example, one can omit sequence ($c1$, $p1$) (see Fig. 8A). However, in this mutant, p fails immediately after performing a c. Hence, the mutant in Fig. 8B models a different and more subtle fault than the mutant in Fig. 8A. Thus, the set of fault models can be extended by generating mutants modeling different or more subtle faults.

Actually, this example hints that using morphologically different models in test generation is also beneficial since it helps the detection of different or more subtle faults.

2.4 Test Generation From Models

Although it is possible to use different coverage criteria for test generation considering different (even application-specific) semantics, such as inter-component coverage criterion [8], which utilizes the structure of the GUIs, we limit our discussion for test generation to more generic k-sequence and faulty k-sequence coverage criteria for detection of missing event and extra event faults as discussed in Section 2.2.

Given an event-based model G and a set of sequences X. X is said to cover a k-sequence r in M, if r appears in a sequence in X, and, if X covers all k-sequences in G, it is said to achieve *k-sequence coverage*. Furthermore, X

A

S → c1 c1 c(c1 c1) | c1 x1 c(c1 x1) | c1 p1 c(c1 p1) | x1 c1 c(x1 c1) | x1 x1 c(x1 x1) | x1 p2 c(x1 p2)
c(c1 c1) → c1 c1 c(c1 c1) | c1 x1 c(c1 x1) | c1 p1 c(c1 p1)
c(c1 x1) → x1 c1 c(x1 c1) | x1 x1 c(x1 x1) | x1 p2 c(x1 p2)
c(c1 p1) → p1 c1 c(p1 c1) | p1 x1 c(p1 x1) | p1 p1 c(p1 p1) | ε
c(x1 c1) → c1 c1 c(c1 c1) | c1 x1 c(c1 x1) | c1 p1 c(c1 p1)
c(x1 x1) → x1 c1 c(x1 c1) | x1 x1 c(x1 x1) | x1 p2 c(x1 p2)
c(x1 p2) → p2 c1 c(p2 c1) | p2 x1 c(p2 x1) | ε
c(p1 c1) → c1 c1 c(c1 c1) | c1 x1 c(c1 x1) | c1 p1 c(c1 p1)
c(p1 x1) → x1 c1 c(x1 c1) | x1 x1 c(x1 x1) | x1 p2 c(x1 p2)
c(p1 p1) → p1 c1 c(p1 c1) | p1 x1 c(p1 x1) | p1 p1 c(p1 p1) | ε
c(p2 c1) → c1 c1 c(c1 c1) | c1 x1 c(c1 x1) | c1 p1 c(c1 p1)
c(p2 x1) → x1 c1 c(x1 c1) | x1 x1 c(x1 x1) | x1 p2 c(x1 p2)

B

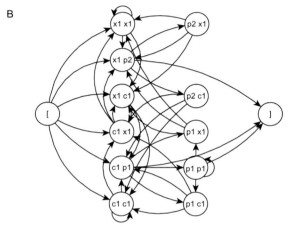

Fig. 7 A 2-Reg model (transformed from Fig. 5) [10]. (A) Productions. (B) Directed graph visualization.

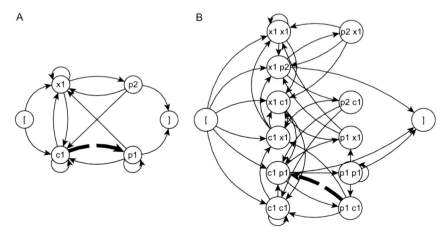

Fig. 8 Two mutants generated from morphologically different models (mutations are shown in *boldface dashed lines*) [10]. (A) A mutant of the model in Fig. 5. (B) A mutant of the model in Fig. 7.

is said to cover a faulty k-sequence r which is not in G, if r appears in a sequence in X, and, if X covers all k-sequences not in G, it is said to achieve *faulty k-sequence coverage*.

As discussed in Section 2.2, k-sequence coverage is used to reveal missing event faults where an event does not follow a (possibly empty) sequence of events. Although k-sequence coverage can be used to reveal different or more subtle faults as k is increased, it is not stronger for increasing value of k; that is, $(k+1)$-sequence coverage does not subsume k-sequence coverage for $k \geq 1$. It is possible that a k-sequence is not included in any $(k+1)$-sequences. In this case, a sequence set achieving m-sequence coverage for $m \geq k+1$ fails to cover such k-sequences. If a complete subsumption is intended, such sequences should be singled out and included separately.

In addition, faulty 1-sequence coverage is different from faulty k-sequence coverage for $k \geq 2$ because the faultiness of an event depends on its preceding event. Faulty 1-sequences actually correspond to faulty start events and, therefore, they should be covered at the beginning of the sequences. In general, faulty k-sequence coverage is used to reveal extra event faults where an event follows a (possibly empty) sequence of events (although it should not). For this reason, we only consider the faulty k-sequences whose last events are faulty.

To satisfy these criteria, we use k-Reg models. However, there is a problem: A sequence s in the corresponding k-Reg G_k of a 1-Reg G_1 is not always a sequence in G_1. Therefore, in order to obtain a sequence t in G_1 from a sequence s in G_k, the following transformation is defined.

Let $s = u^1 \ldots u^m$ where $k \geq 1$, $m \geq 1$, and $u^i = u^i{}_1 \ldots u^i{}_k$ for $i = 1, \ldots, m$. *Inverse sequence transformation of s based on integer k is a* $(k+m-1)$*-sequence*

$$T_S^{-1}(s, k) = u^1\, u^2{}_k\, u^3{}_k \ldots u^m{}_k$$

where $u^1 = s_1 \ldots s_k$ and each $u^i{}_k = s_{i \times k}$ for $i = 2, \ldots, m$.

For example, $s = c1\, c1\, c1\, x1\, x1\, x1\, x1\, p2$ is a sequence in the 2-Reg in Fig. 7, but it is not in the 1-Reg in Fig. 5. However, $T_S^{-1}(s, 2) = c1\, c1\, x1\, x1\, p2$ is a 5-derived sequence in the 1-Reg in Fig. 5.

2.4.1 Positive Test Generation

In order to generate test cases achieving $(k+1)$-sequence coverage from a given 1-Reg, its corresponding k-Reg can be used. In this way, one can reveal missing event faults where an event does not follow a certain k-sequence.

As discussed in Section 2.1, productions of a k-Reg encode $(k+1)$-sequences in the system. Hence, by covering these productions, one can generate test sets achieving $(k+1)$-sequence coverage.

Algorithm 4. Test Generation to Achieve $(k+1)$-Sequence Coverage

```
Input: G = (E, B, K, C, S, P) - the input 1-Reg
       k - an integer ≥ 1
Output: X - a set of sequences which achieves (k+1)-sequence
        coverage for G
X = ∅
Gₖ = transform G to its corresponding k-Reg   //See Algorithm 3
     in Section 2.3
Y = generate a sequence set achieving production coverage for Gₖ
for each s ∈ Y such that |s| ≥ 2k do
    X = X ∪ Tₛ⁻¹(s, k)   //See Section 2.4
endfor
```

Algorithm 4 outlines the generation of a test set achieving $(k+1)$-sequence coverage from a given 1-Reg. In the algorithm, generating a test set achieving production coverage for the corresponding k-Reg model can be performed in different ways. For example, it is possible to perform some optimizations by adapting algorithms to solve Chinese Postman Problem over directed graphs, like Refs. [35–37], to cover each production a minimum number of times, resulting in a reduced set of test cases. However, one should note that optimization algorithms tend to require more resources in terms of both time and space, and there is no guarantee of reduced test execution costs [20,38]. Thus, algorithms such as those in Refs. [39–41] can also be used to generate relatively short but generally nonoptimized sequences from a given grammar, while using less resources.

When Algorithm 4 is executed on the 1-Reg in Fig. 5 for $k=1$, no transformation of the grammar is necessary. One can obtain the following set of test cases

$$\{c1\,c1\,x1\,c1\,p1\,c1\,p1\,x1\,x1\,p2\,c1\,p1\,p1\,x1\,p2\,x1\,p2\,c1\,p1\}$$

which achieves 2-sequence coverage. Furthermore, if $k=2$ is used, the given 1-Reg is transformed once to obtain the 2-Reg in Fig. 7, this 2-Reg is used to generate a sequence set and the elements of this set are inverse transformed to obtain test cases achieving 3-sequence coverage. The following is an example of test cases achieving 3-sequence coverage:

$$\{c1\ c1\ c1\ x1\ c1\ c1\ p1\ c1\ c1\ p1\ x1\ c1\ x1\ x1\ c1\ p1\ p1\ c1\ x1\ p2\ c1\ c1\ p1,$$
$$c1\ x1\ p2\ x1\ c1\ p1\ c1\ p1\ x1\ x1\ x1\ p2,$$
$$c1\ p1\ x1\ p2\ c1\ x1\ p2\ c1\ p1\ p1\ x1\ p2\ x1\ x1\ p2\ x1\ p2,$$
$$x1\ c1\ p1\ p1\ p1\ x1\ x1\ p2\ c1\ p1\ p1\}.$$

Naturally, during test execution, the corresponding basis event is used for each event, because basis events represent the events as they are visible to user.

2.4.2 Negative Test Generation

In order to generate test sets achieving fault $(k+1)$-sequence coverage, certain k-Reg mutants need to be generated and used. Therefore, before laying out the test generation algorithm, we discuss two mutant selection strategies based on mark start and insert k-sequence mutants. (A detailed discussion on other possible k-Reg mutants and on the reasons for the selection of only these two types of mutants can be found in Ref. [33].) Having discussed mutant-related background, we can now proceed to the test generation algorithm to achieve faulty $(k+1)$-sequence coverage as demonstrated in Algorithm 5.

Algorithm 5. Test Generation to Achieve Faulty $(k+1)$-Sequence Coverage

```
Input: G = (E, B, K, C, S, P) - the input 1-Reg
       k - an integer ≥ 1
Output: X - a set of sequences which achieves single-end faulty
            (k+1)-sequence coverage for G
X = ∅, Y = ∅
G_k = transform G to its corresponding k-Reg
    //See Algorithm 3 in Section 2.3
for each G' = Ms(G_k, e) selected using Algorithm 1
        in Section 2.2.3 do
    X = X ∪ {e_1}
endfor
for each G' = It(G_k, a, e) selected using Algorithm 2
        in Section 2.2.3 do
    s = generate a sequence ending with e by covering production
        c(a) → e c(e) from G'
    X = X ∪ T^-1(s, k)    //See Section 2.4
endfor
```

When Algorithm 5 is executed on the 1-Reg in Fig. 5 for $k=1$, one can obtain the following test set:

$$\{p1, x1\,p2\,p3\}$$

Furthermore, if $k=3$ is used, the given 1-Reg is transformed twice to obtain the corresponding 3-Reg, the mutants of this 3-Reg are used to obtain test cases. The following is an example test set:

$$\{p1, c1\,x1\,p2\,p3, x1\,x1\,p2\,p3, c1\,p1\,x1\,p2\,p3, x1\,p2\,x1\,p2\,p3\}$$

As usual, the corresponding basis event is used for each event during test execution.

2.5 Issues to Consider

There are certain issues related to the model-based GUI testing methodologies discussed so far in Section 2 which are worth mentioning.

Model Semantics: The models that are most commonly used for GUI testing have very simple semantics. They are mainly based on follows relation between events. This kind of abstractions may not be sufficient to capture the relevant behavior of certain applications.

Fault and Coverage Semantics: Being partially related to the semantics of the model, fault models and coverage criteria are relatively general. In case one is interested in only a specific fault type, the employed fault models and coverage criteria may be more than needed causing a waste of resources.

Size Complexity of Morphology Variation: In theory, the proposed morphology variation technique causes an exponential increase in the model size. Although, in practice, the value of k is almost always bounded, it still poses a limit by preventing the use of relatively larger k values.

Linear Test Cases: In most cases, generated test cases have a linear structure and they are composed of input events. This does not allow the possibility of a change in the flow of execution depending on responses from the system.

Order of Test Cases: Algorithms to generate test cases do not impose a specific order on the test cases. However, in practice, the fault detection efficiency of an approach may change depending on the order of executed test cases.

3. TESTING AND TEST OPTIMIZATION EXEMPLIFIED BY GUI-MODELING WITH ESG

Model-based GUI testing establishes a model that is used for guiding the test process to generate and select test cases, which form *sets* of test cases, or test *suites*. The selection is ruled by an *adequacy criterion,* which provides a measure of how effective a given set of test cases are in terms of its potential to reveal faults. Most model-based approaches use *coverage-oriented* adequacy criteria which determine how well the generated test suites cover the corresponding model. The ratio of the portion of the specification or code that is covered by the given test set in relation to the uncovered portion can then be used as a decisive factor in determining the point in time at which to stop testing (*test termination*).

Memon *et al.* introduced an approach to testing GUI [42]. It deploys methods of knowledge engineering to generate test cases, test oracles, etc., and to deal with the test termination problem. The approach uses some heuristic methods to cope with the state explosion problem, which has the disadvantage that there is no guarantee that the best solution found is the optimal solution.

Marré and Bertolino [43] adopted the notion of "spanning set," which is similar to what has been introduced as "minimal spanning set of complete test sequences." Gargantini and Heitmeyer used a state-oriented approach [44], which is based on the traditional SCR (Software Cost Reduction) method. The approach uses model checking to generate test cases automatically from formal requirement specifications, using coverage metrics for test case selection. The approach is limited by the state space explosion problem.

Using the results of aforementioned approaches, and based on ESG notation (Section 2), this section introduces a different method for GUI test optimization [45].

3.1 Test Termination as an Optimizing Problem

Coverage-oriented test termination must be, of course, economical in terms of an efficient test suite that is generated by optimizing the test execution time. Test execution time accounts for a significant portion of the overall test execution costs. Tests that are generated according to the coverage adequacy criteria are mostly too expensive to be executed as they require longer execution time, which leads to an efficiency deficit. Test sets need to be specifically structured to optimize the test execution time. The number and length of test cases of a test set are the primary factors that influence the cost of the test execution time in an automated test framework which does not need further human intervention or effort [45].

3.1.1 Minimizing the Test Sets of ESGs

As defined in Section 2, subsequent nodes traversed through an ESG represent an event sequence (ES). Sequences of length 2 are called *event pairs*

(EPs). An event sequence is *complete* (CES) if the sequence includes the start and finish node and is also called a *walk* in the following. The union of the sets of CESs of minimal total length to cover the ESs of a required length is called *Minimal Spanning Set of Complete Event Sequences* (MSCES). If a CES contains all EPs at least once, it is called an *entire walk*. A legal entire walk is minimal if its length cannot be reduced. A minimal legal walk is *ideal* if it contains all EPs exactly once. Legal walks can easily be generated for a given ESG as CESs, respectively. It is not, however, always feasible to construct an entire walk or an ideal walk. Using some results of the graph theory [46], MSCESs can be constructed as the next section illustrates.

3.1.2 Minimal Spanning Sets of Complete Event Sequences

As mentioned in Section 2, a CES represents a *legal* walk, traversing the ESG from its entry to the exit. Given an ESG e, a complete legal walk contains each EP in e at least once. A complete legal walk is *minimal* if its length cannot be reduced without changing it to an incomplete legal walk. A minimal legal walk is considered *ideal* when it contains every EP exactly once. Legal walks can be generated easily for a given ESG as CESs. It is not, however, always feasible to construct a complete or an ideal walk. Using results from graph theory [46], MSCESs can be constructed as follows:

- Check whether an ideal walk exists.
- If not, check whether a complete walk exists and, if so, construct a minimal one.
- If there is no complete walk, construct a set of walks such that (a) sum of the lengths of all walks is minimal, and (b) all EPs are covered.

The MSCES problem introduced here has a lower degree of runtime complexity than the *Chinese Postman Problem* as the edges of the ESG are not weighted, i.e., the adjacent nodes are equidistant. In the following, we summarize results relevant to the calculation of test costs that make the test process scalable. An algorithm described in Ref. [47] to solve the CPP determines a minimal tour that covers the edges of a given strongly connected graph. Transformation of an ESG into a strongly connected graph is illustrated in Fig. 9. Addition of a backward edge, indicated as a dashed

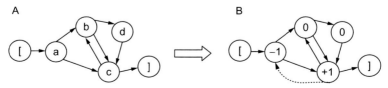

Fig. 9 (A) An example ESG. (B) Transferring walks into tours and balancing the node.

arrow from the exit to the entry, transforms the ESG in Fig. 9A to a strongly connected graph in Fig. 9B.

The labels of the vertices in Fig. 9B indicate the *balance* of these vertices as the difference between the number of incoming edges and the number of the outgoing edges. These balance values determine the number of additional edges that will be identified by searching all shortest paths and solving the optimization problem. The problem can then be transformed into the construction of an *Euler tour* for this graph [46]. This tour may have multiple occurrences of the backward edge indicating the number of walks. For the ESG in Fig. 9B, based on Fig. 9A, the minimal set of the legal walks covering the EPs are **MSCES** = {*abcbdc, ac*}. Note that no complete walks exist. Therefore, an ideal walk cannot be constructed.

Algorithm 6 calculates the MSCES for a given ESG as input. ε denotes the entry of the ESG and γ its exit. Given an event $v \in V$, diff(v) denotes the number of predecessor events of v minus the number of its successor events, which enables the construction of the *bags* (or *multisets*) A, B in the FOR-loop. We introduce the notations $[\![\,]\!]$ for *bags* and $\uplus$ *bag union*. They can be defined informally as follows. For instance, if diff(v) = 3 in the first iteration step, assuming that A is initially empty, the bag A will consist of three instances of v, i.e., A = $[\![v,\ v,\ v]\!]$ after the assignment there. Note that $[\![v,\ v,\ v]\!] \neq \{v\}$, because the two entities on either side of the inequality sign $\neq$ are of different types; on the left-hand side is a bag (with three instances of v), whereas on the right-hand side is a singleton set with one element v. Turning to $\uplus$, note that $[\![v,\ v,\ v]\!] \uplus [\![v]\!] = [\![v,\ v,\ v,\ v]\!]$.

Algorithm 6. Generation of MSCES [48]

    Input: ESG = (V, E, Ξ, Γ); ε =[, γ =]

    Output: MSCES

```
add_arc(ESG, (γ, ε));
bags A, B, M = [[]]; set MSCES = ∅;                    //empty bags & set
FOR all nodes v∈V DO
    IF (diff(v) > 0) THEN FOR i:=1 TO diff(v) DO A = A ⊎ [[v]];
    IF (diff(v) < 0) THEN FOR i:=1 TO diff(v) DO B = B ⊎ [[v]];
m = |A| = |B|;
        //cardinality
D[1 .. m][1 .. m];
        //distance matrix D
FOR all nodes v∈A DO
    compute_shortest_paths(v, B, D);
```

```
M = solveAssignmentProblem(D);
FOR all (i, j)∈M DO
   Path = get_shortest_path(i, j);
   FOR all arcs e∈Path DO
      add_arc(ESG, e);
EulerTourList = compute_Euler_tour(ESG);
start = 1;
FOR i=2 TO length(EulerTourList)-1
   IF (getElement(EulerTourList, i)= γ) THEN
      MSCES = MSCES ∪ getPartialList(EulerTourList, start, i);
   start = i + 1;
RETURN MSCES;
```

3.1.3 Minimal Spanning Set of FCESs

The concatenation of an event sequence ES and a faulty event pair (FEP), i.e., an event pair of the inverse of complementary ESG, is defined as an FES. An FES is complete (FCES=faulty complete event sequence) if the sequence starts from the entry node. The union of the sets of FCESs of the minimal total length to cover the FESs of a required length is called *Minimal Spanning Set of Faulty Complete Event Sequences* (MSFCES).

In comparison to the interpretation of the CESs as legal walks, illegal walks are realized by FCESs that never reach the exit. An illegal walk is minimal if its starter cannot be shortened. Assuming that an ESG has n nodes and d arcs as EPs to generate the CESs, then at most $u:=n^2-d$ FCESs of minimal length, i.e., of length 2, are available. Accordingly, the maximal length of an FCES can be n; those are subsequences of CESs without their last event that will be replaced by an FEP. Therefore, the number of FCESs is precisely determined by the number of FEPs. FEPs that represent FCES are of constant length 2; thus, they also cannot be shortened. It remains to be noticed that only the starters of the remaining FEPs can be minimized, e.g., using the algorithm given in Ref. [49].

The minimal set of the illegal walks (MSFCES) for the ESG in Fig. 10: *aa, ad, abb, aba, aca, acc, acd, abdb, abdd,* and *abda.*

3.2 Exploiting the Structural Features of SUT for Further Reduction of Test Effort

The approach has been applied to the testing and analysis of the GUIs of different kinds of systems, leading to a considerable amount of practical experience. A great deal of test effort could be saved considering the

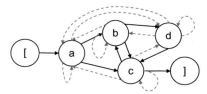

Fig. 10 Completion ESG of Fig. 5A to determine MSFCES.

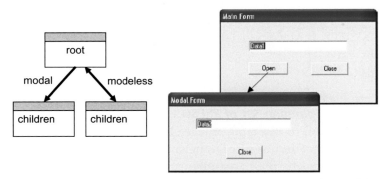

Fig. 11 Modal windows vs modeless windows and an example of a modal opened window.

structural features of the SUT. Thus, there is further potential for the reduction of the cost of the test process [50].

Analysis of the structure of the GUIs delivers the following features:

- Windows of commercial systems are nowadays mostly hierarchically structured, i.e., the root window invokes children windows that can invoke further (grand) children, etc.
- Some children windows can exist simultaneously with their siblings and parents; they will be called *modeless* (or *nonmodal*) windows. Other children, however, must "die," i.e., close, in order to resume their parents (*modal* windows).

Fig. 11 represents these window types as a "family tree." In this tree, a unidirectional edge indicates a modal parent–child relationship. A bidirectional edge indicates a modeless one.

Because modal windows must be closed before any other window can be invoked, it is not necessary to consider the FESs of the parent and children. This is true only for the FCESs and MSFCES as test inputs considering the structure information might impact the structure of the ESG, but not the number of the CESs and MSCESs as test inputs.

Thus, similar to the strong connectedness and symmetrical features [51], the modality feature is extremely important for testing since it avoids unnecessary test efforts.

3.3 Case Studies and Their Empirical Evaluation for the Practice

The objective of the case studies in this section is to determine the increased test effort that arises in relation to the length/number of ESs to be covered and to find out whether this additional test effort is rewarded adequately by the revelation of additional errors. The data needed for this analysis were collected and evaluated by means of experiments carried out in accordance with the principles of software experimentation [52]. Case study 1 focuses on the test effectiveness detecting defects by the various coverage types with different lengths [53]. The second case study, case study 2, is focusing on the test efficiency, i.e., the test cost reduction achieved from minimal test sequences [45].

3.3.1 Case Study 1: Software Application GUI Under Test

For the case study, RealJukebox (RJB) has been selected, more precisely the basic, English version of the RJB 2 (Build: 1.0.2.340) of RealNetworks. There are several reasons why RJB has been selected to be SUT. First, RJB as SUT is a commercial, popular application that is widely well known and accepted by a great variety of users. Second, the selected SUT has been be used over many years in different languages and in cultural contexts. Furthermore, RJB has been frequently updated and, therefore, is mature and well established. Last but not least, RJB makes comfortable use of dynamic window components in several hierarchy levels. The basic configuration of the tested RJB consists of about 200 distinct components. To sum up, choosing the RJB as SUT avoided studying an "alpha" version of a no-name product for the case study with the present approach.

3.3.1.1 Results and Analysis

Table 1 arbitrarily extracts some of the detected faults. The fault reproduction process is very simple. As an example, in order to reproduce the fault No. 1, one starts with the Control option of the Main Menu of the RJB (see Fig. 12) and subsequentially pushes the Rec button and then FF button. In Fig. 12, the dashed line with the label No. 1 uniquely identifies this sequence of actions. The other faults of Table 1 can be reproduced the same way.

Table 1 Excerpt From the List of Faults Revealed by Testing the System Function "Play and Record a CD or Track"

No	Detected Faults	Test Case
1	While recording, pushing the forward button or rewind button stops the recording process without a due warning	*Record FF*
2	If a track is selected but the pointer refers to another track, pushing the play button invokes playing the selected track; i.e., the situation is ambiguous	*Select Track Play*
3	Menu item Play/Pause does not lead to the same effect as the control buttons that will be sequentially displayed and pushed via the main window. Therefore, pushing play on the control panel while the track is playing stops the playing	*Play Play*
4	Track position could not be set before starting the play of the file	*Trackposition Play*
5	Record Shuffle does not activate shuffling, i.e., tracks will be processed sequentially	*CheckOne ++ Shuffle Record*
6	If the track is in Pause and Record button is pushed, then the track will be played	*Play/Pause Play/ Pause Record*
7	The system jumps to a track that was not selected and terminates the playback although the selected tracks have not been completely played	*Play/Pause FF FF FF*

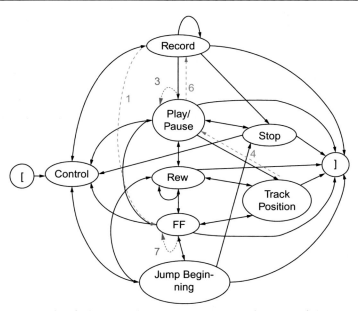

Fig. 12 FEP revealed faults in a subgraph representing "Playing track."

Note that the fault Nos. 2 and 5 are not included in Fig. 12 as they are detected via other ESGs. Due to lack of space, these completed ESGs are not included in this chapter.

Table 2 summarizes the number of test cases, their length, and the corresponding faults. Faults are further classified as surprises and defects. *Defects* are serious departures from specified behavior; *surprises* are user-recognized departures from expected behavior. A surprise behavior is not explicitly indicated in the specification of the UI; it should, however, be perceived by some users as a disturbing or disappointing behavior of the system.

The number of defects detected by test cases of lengths 3 and 4 increases obviously slower in relation to those of length 2. Since the faults are independent, these longer tests should still be executed, if the test budget and time allow for this. Another reason why test cases of lengths 3 and 4 should

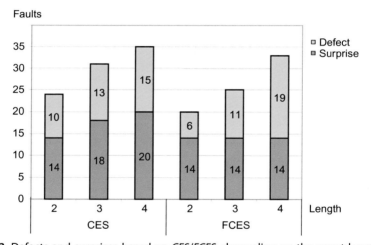

Fig. 13 Defects and surprises based on *CES/FCES*, depending on the event length.

Table 2 Test Case Costs and Detected Faults Depending on the Event Length

Length of Covered ES	Test Cases	Detected Faults by CES	Detected Faults by FCES	Total No. of Detected Faults	Surprises	Defects	Fault per Test Case (Efficiency)
2	914	24	20	44	16	28	4.8×10^{-2}
3	2458	24 + 7	20 + 5	56	16 + 8	28 + 4	2.3×10^{-2}
4	6936	31 + 4	25 + 8	68	24 + 10	32 + 2	0.9×10^{-2}

be executed is given by the likely severity of the "expensive" faults, i.e., defects that can only be detected with these longer, thus more "expensive," tests. This situation is simple to explain: The longer the test procedure lasts, the less populated the remaining faults become, while one might expect to detect more intricate and subtle faults.

Fig. 13 depicts the results found. It can be observed that the tests based on the CESs of length 4 and on FCESs of length 4 are very beneficial in detecting defects: 19 defects have been detected by tests based on FCES of length 4 in relation to only 6 based on FCESs of length 2! Thus, a clear tendency can be observed that an increasing number and length of CES-based and FCES-based test cases lead to the detection of an increasing number of defects. Note, however, that Fig. 13 does not consider the number of necessary tests, i.e., test costs.

3.3.1.2 Lessons Learned
Now, summarizing the observations of the test process when performing the case study and their implications lead to following major findings:
- The RJB that has been tested in this section is a product that has been matured in many years of intensive and extensive deployment.
- The fact that so many faults could be detected in this product motivates the refinement and improvement of this approach.

To sum up further results, Table 2 and Fig. 13 clearly show the fact that CES-based tests of length 2 are the most cost-effective ones in this approach, i.e., they detect faults at the lowest costs per fault. In other words, there is a rapid fall off in cost-effectiveness of length of the event sequences to be covered as a consequence of the rapid rise of the number of the test cases (a total of 914 tests to cover ESs of length 2 for all 12 functions, whereas 6936 tests to cover their ESs of length 4; cf. Table 2). This finding can be explained by analyzing the structure of the SUT: The number of CESs to cover ESs that are longer than 2 primarily increases with the number of loops within the ESGs. In other words, the more vertices of the ESG under consideration are connected with each other, the larger is the number of tests to cover ESs that are longer than 2.

On the other hand, tests that cover ESs of lengths 3 and 4 seem to detect more and intrinsic faults (if any), however at considerably more cost per detected fault. Finally, CES-based tests are more cost efficient than the ones which are FCES based.

Based on these observations, it is strongly recommended starting the test process with the CES-based test cases of length 2, further continuing with the CES-based test cases of lengths 3 and 4 and finally to start with the FCES-based ones. If the cumulative number of the detected faults grows very slowly, one can terminate the test process after execution of the CES-/FCES-based test cases to cover the ESs of length 4 as further tests become very cost ineffective. This really cannot be considered as a tendency of "reliability growth" because the detected faults were not corrected; they have just been ignored; however, the multiple counting of the same faults has been avoided.

The approach delivers a very simple, but nevertheless a cost-effective, stepwise, and straightforward test strategy, because the approach enables the enumeration of the test cases and, consequently, the scalability of the test process.

3.3.2 Case Study 2: Embedded Software GUI Under Test
The SUT we use in the examples is a control terminal of a marginal strip mower (Fig. 14) which controls a marginal strip mower (RSM13) of a special, heavy-duty vehicle (Unimog of Mercedes-Benz). This display unit takes the optimum advantage of mowing around guide poles, road signs, and trees. Operation is effected either by the power hydraulic of a light truck or by the front power takeoff. Further buttons on the control desk simplify the operation, so that, e.g., the mow head returns to working position or to transport position when a button is pressed.

3.3.2.1 Results and Analysis
For a comprehensive testing, several strategies have been developed with varying characteristics of the test inputs, i.e., stepwise and scalable increasing and/or changing the length and number of the test sequences, and the type of the test sequences, i.e., CES- and FCESs-based, and their combinations.

Fig. 14 The example vehicle.

Table 3 Two of the Detected Faults of the RSM Control Terminal

No.	Faults Detected by the FCES
1.	The cutting unit can be activated without having any pressure on the bottom, which is very dangerous if pedestrians approach the working area
2.	Keeping the button for shifting the mow head pushed and changing to another screen causes control problems of shifting: The mower head with the cutting unit cannot immediately be stopped in an emergency case

Table 4 Reducing the Number of Test Cases

Length	#CES	# MSCES	Cost Reduction ES (%)
2	40	15	62.5
3	183	62	66.1
4	549	181	67.0
Sum	772	258	65.2

Length	# MSFCES Without Structural Information	# MSFCES With Structural Information	Cost Reduction MSFCES (%)
2	75	58	22.7
3	167	218	35.7
4	487	292	40.0
Sum	729	568	32.8

Following could be observed: The test cases of the length 4 were more effective in revealing dynamic, intricate faults than the test cases of the lengths 2 and 3. Even though more expensive to be constructed and exercised, they are more efficient in terms of costs per detected fault. Further on the CES-based test cases as well as the FCES-based cases were effective in detecting faults.

Due to the lack of space, the experiences with the approach are here very briefly summarized. This can be, however, found in Ref. [53]. To sum up the test process, one student tester carried out 826 tests semiautomatically and detected a total of 39 faults, including some severe ones (Table 3).

In a second stage, the results of the research work for minimizing the spanning set of the test cases (MSCES and MSFCES) have been applied to the testing of the margin strip mower. Table 4 demonstrates that the minimization algorithm (Section 3.1.1) could save in average about 65% of the total test costs, while the exploitation of the structural information (Section 3.2) of the SUT could further save up to almost 30%.

3.3.2.2 Lessons Learned

3.3.2.2.1 Lesson 1. Start Small, but as Early as Possible
The determination and specification of the CESs and FCESs should ideally be carried out during the definition of the user requirements, much before the system is implemented; the availability of a prototype would be helpful in this task. They are then a part of the system and the test specification. However, CESs and FCESs can also be produced incrementally at a later time, even during the test stage, in order to discipline the test process.

As a strategy, one starts with the CESs and FCESs that cover all event pairs. Test results and quality targets determine how to proceed further, i.e., whether to consider testing with event triples and quadruples.

3.3.2.2.2 Lesson 2. Good Exception Handling Is Not Necessarily Expensive but Rare
Most GUIs subjected to tests do not consider the handling of the faulty events. They have only a rudimentary, if any, exception handling mechanism, realized by a "panic mode" that mostly leads to a crash or ignores the faulty events. The number of the exceptions that should be handled systematically but have not been considered at all by the GUIs of the commercial systems is presumed to be on an average about 80%.

3.3.2.2.3 Lesson 3. Analysis Prior to Testing Can Reveal Conceptual Flaws
The analysis of ESGs of the GUIs of some commercial systems has revealed several conceptual flaws: absence of edges, indicating incomplete exception handling, and missing vertices or events (approximately 20%). This amounts to defective components in the final product, highlighting the flaws in the initial concept and the process of product development. In this connection, the proposed approach offers an important unexpected benefit: It provides a framework for the accelerated maturation of the product and for exercising the creativity of the developers.

4. CONTRACT-BASED TESTING OF GUIs

Methods and techniques presented in Sections 2 and 3 aim to generate abstract test cases. In this section, a contract-based approach for concrete test case generation is presented. The presented contract-based approach stems from the concept of design by contract (DbC). Meyer [54] introduced DbC as an object–oriented design technique. Design by contract follows the principle that interfaces between modules of a software system should be governed by precise specifications, similar to contracts between humans or companies. The contracts cover mutual obligations (preconditions), benefits (postconditions), and consistency constraints (invariants) [55]. From the point of view of GUI testing, DbC plays an important role because contracts delineate what the user is expected to provide as input and what the GUI is expected to supply as output with respect to the provided input. From the testing point of view, a GUI operation can be evaluated with respect to preconditions, postconditions, and invariants according to DbC. Following DbC, UML is supplemented with object constraint language to provide some formalism. Contracts in decision table format are compact and easy to understand and maintain [56]. However, there is no generally accepted formalism for contract representation.

4.1 Contract-Based Testing in General

Contracts form a valuable source of information regarding the intended semantics of the software. A behavioral specification is a description of what is expected to happen when software executes [57]. This specification can be used to verify that the software meets its requirements. When behavioral specification is presented using DbC, it becomes more useful for both programmers and testers.

There exist some approaches that adopt the DbC–idea for testing. Zheng and Bundell [58] introduced an UML-based software component testing technique called Test by Contract. Ciupa and Leitner [59] noted that the validity of a software element can be ascertained by checking the software with respect to its contracts. In addition to using contracts to automatically generate test input values, contracts can be used as test oracles as they define valid and invalid conditions for the software. Thus, utilization of contracts eliminates the necessity of developing a test oracle for each test case [60]. As contracts are used to evaluate test results, the quality of the test oracles is entirely dependent on the quality of the contracts [59]. Aichernig [61] shows how mutation testing can be applied to contracts. Similar to source code mutation, a mutant contract is produced by introducing small change to the formal contract definition. Then test cases that are able to detect the introduced mutations are selected.

Madsen [62] investigated how JUnit and Design by Contract can be combined. This way, assertions written as pre- and postconditions and class invariants can be used in unit test cases and automatic execution and evaluation of test cases become possible. If test cases are automatically generated, then JUnit can be used to set up and execute the test cases and then contracts can be used to evaluate the test cases. Languages like Phyton, C++, and Java are extended to comply with DbC for catching bugs. Guerreiro [63] used the design by contract in C++ by using and inheriting the Assertions class. In Ref. [64], the DbC concept is integrated into the programming language Python and adopted by adding mechanisms for dynamic type checking of method parameters and instance variables.

Contracts establish the ground for the automation of the MBT process. While testing a system, a model of the system helps to predict and control its behavior. Modeling a system acquires the understanding of its abstraction, and there is the need of a formal specification technique for distinguishing between legal and illegal situations. Contracts serve perfectly for these purposes. There are some contract-based testing techniques focused on web service testing [65,66], where web service behavior is modeled using contracts. Valentini et al. [67] proposed a framework for contract-based component testing, which enables extendable and robust contract checkers to be dynamically inserted between client component and supplier component. Contract checkers use the contract between client and supplier to work like proxies by forwarding method call to the client, the result back to the supplier and to evaluate the test result. Xu et al. [68] proposed an approach that transforms a contract-based test model into an operational model, which enables analysis of the correctness of the test model. Then integration tests are generated to meet coverage criteria of the test model.

Another use of contracts is for robustness. Robustness is a quality attribute, which is defined by the IEEE standard glossary of software engineering terminology [69] as the degree to which a system or a component can function correctly in the presence of invalid inputs or stressful environmental conditions. Specifications presented in contracts helps to improve the robustness of software [70]. An example of contract usage in robustness testing is presented by Tuglular et al. [71], where they introduced decision table augmented ESGs, which utilizes the design by contract patterns, and applied this concept to event-based robustness testing for catching boundary overflow errors.

4.2 ESG-Based Contract Testing of GUIs

The contract notion is used to describe input properties in precise terms. Preventing invalid input from ever getting to the application in the first place is possible only at the UI. Therefore, GUIs should be specifically designed to filter unwanted or unexpected input. This can be achieved through input contracts that are defined and used in our work. Model-based specification of input contracts is achieved through an input contract model, which enables the input data and corresponding actions to be defined with their constraints. Thus, for simplicity, the term "testing" here is used to refer to function-based testing, specification-oriented testing, or black-box testing.

In the input contract testing approach, the tests are derived from contracts supporting the creation of test input values and test oracles. This novel approach suggests that an automatic input testing process is possible with a GUI test driver that invokes mouse clicks and enters text into rich client GUIs. In this context, contracts form a valuable source of information regarding the intended semantics of the software. As noted by Ciupa and Leitner [59], the validity of a software element can be ascertained by checking the software with respect to its contracts. Therefore, contracts establish the ground for the automation of the testing process. Accordingly, the primary goal of input contract testing is to develop and implement a fully automated test case generation for contract-based GUI input testing.

The input contract testing approach suggests converting GUI specification into a model, which is employed to generate positive and negative test cases. The ESG is chosen for the specification of GUIs. ESG merges inputs and events and turns them to vertices of an event transition diagram for easy understanding and checking the behavior of the GUI under consideration.

4.2.1 Input Contract Model

Modeling input data, especially concerning causal dependencies between each other as additional nodes, inflates the ESG model since vertices represent events and edges allowed sequences of events and not transitions as in automata theory. Assuming that a condition for choosing input data can be evaluated to true or false, the combination of conditions results in $2^{|C|}$ combinations, where $|C|$ represents the number of conditions. Each combination of conditions would have to be modeled as vertex and is to be connected with the appropriate successor. Thus a decision table (DT) with n binary conditions subsumes 2^n nodes to realize a thorough evaluation considering all combinations. To avoid this inflation, decision tables are introduced to refine a node of the ESG. Such refined nodes are double circled. The successors of such refined vertices represent the actions of the DT and vice versa.

A *Decision Table* $DT = (C, A, R)$ represents actions that depend on certain constraints, where

- $C \neq \emptyset$ is the set of *constraints* (*conditions*) as Boolean predicates,
- $A \neq \emptyset$ is the set of *actions*, and
- $R \neq \emptyset$ is the set of *rules*, each of which triggers executable actions depending on a certain combination of constraints.

Decision tables are popular in information processing and are used for testing, e.g., in cause and effect graphs. A DT logically links conditions ("if") with actions ("then") that are to be triggered, depending on combinations of conditions ("rules").

Let R be defined as the set of rules, each of which triggers executable actions depending on a certain combination of constraints. Then, a *rule* $r \in R$ can be defined by

$$r = (C_{True}, C_{False}, A_x),$$

where
- $C_{True} \subseteq C$ is the set of constraints that have to be resolved to true,
- $C_{False} \subseteq C$ is the set of constraints that have to be resolved to false, and
- $A_x \subseteq A_{ui} \times A_{xcpt}$ is the set of actions that should be executable if all constraints $t \in C_{True}$ are resolved to true and all constraints $f \in C_{False}$ are resolved to false with
  - A_{ui} containing possible user interactions,
  - A_{xcpt} containing exception messages.

That is, one rule represents a specific combination of conditions where each condition is evaluated either to true or to false. Depending on one rule, one or several follow-on actions are allowed. In the other way around, the execution of a specific action is only allowed if input data is chosen along a rule which possesses the considered action as allowed successor. As already stated earlier, the combination of conditions results in $2^{|C|}$ combinations, that is, $2^{|C|}$ rules can be formulated without producing redundancy. Note that $C_{True} \cup C_{False} = C$ and $C_{True} \cap C_{False} = \emptyset$ under regular circumstances. In certain cases, it is inevitable to mark conditions with a *don't care* (symbolized with a "-" in DT), i.e., such a condition is not considered in a rule and $C_{True} \cup C_{False} \subset C$. A DT is used to refine data input of GUIs.

An example of DT is given in Fig. 15. This DT can be used to refine a node of an ESG. This node will be double circled and next event, which is an action in the DT, is decided with respect to DT that is attached to this double-circled node. Such an ESG is called DT-supplemented ESG and is shown in Fig. 15.

For DTs, such as the one presented in Fig. 15, X entry indicates an action, or for GUIs a user interaction. No exception is defined for actions y and z. As an example, rule 1 (R_1) reads as follows: **If** v_0 is resolved to *true* and v_1 is resolved to *false*, **then** action y will be executed. If this DT is used to refine a node of ESG, such as given in Fig. 15, then regarding to R_1 next event after v will be y and the ES will be $(\ldots, v, y, \ldots)$.

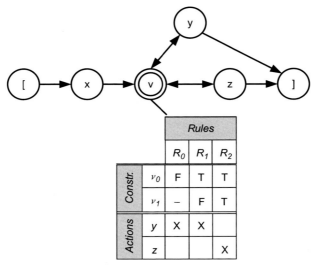

Fig. 15 An example of DT-supplemented ESG [72].

Given a GUI, the quadruple σ is an *input contract model* (Io, Dv, Ac, Co), where

- Io is the finite set of *GUI input objects*;
- Dv is the finite set of data variables;
- Ac is the finite set of actions on GUI;
- Co is the input contract definition represented by a DT.

As stated before, event sequence of GUI is modeled with DT-supplemented ESG. It is an ESG with a special DT, where conditions of DT come from constraints of input contracts. Input contract model provides a guideline for the construction of a contract–supplemented ESG for the GUI it represents. Input objects, such as inputArea and comboBox, and button objects indicate possible events. Event sequences are established among these events through drawing edges between vertices.

4.2.2 Input Contract Testing Process

For GUI input contract testing, test scope is always a GUI. A set of GUI components that make up a window can be tested if event-based testing is integrated into input contract testing. Therefore, GUI input contracts are modeled with contract-supplemented ESGs so that a seamless testing process can be achieved for a window or a composition of GUI input elements. Solutions for automating test case generation and test result interpretation stages are described in the following paragraphs.

A *test case* specifies input values for a method of an input component, which may work on one or more input area. A *test suite* is composed of test cases to check the validation of all assertions offered by an input contract. The input values making up a test case can be derived from the constraints of a provided contract. Expected outputs are actions with or without exceptions given in DT. Please note that an input contract is not supposed to cover all inputs, its purpose is to filter.

GUI input contract testing process is given in Algorithm 7. Full event coverage and full rule coverage criterion is fulfilled in terms of coverage. For full event coverage criterion, each event is executed at least once. In other words, each node of ESG is visited at least once. For full rule coverage criterion, each rule should be tested independently. These test cases should be sampled from input space composed of valid and invalid values of constraints.

Algorithm 7. GUI Input Contract Testing Process

```
generate the corresponding ESG
cover all events by means of CESs
foreach CES with decision tables do
    generate data-expanded CES using corresponding DT (input contract-
        based test case generation)
apply the test suite to GUI
observe GUI output to determine whether a correct response or a faulty
event occurs
```

4.2.3 Input Contract Test Generation

The DT is used to produce test cases automatically. Since it is often not feasible to include all possible input values for a test case, the central question of testing is about the selection of test input values most likely to reveal faults. This problem comes down to grouping data into equivalence classes, which should comply with the property that if one value in the set causes a failure, then all other values in the set will cause the same failure. Conversely, if a value in the set does not cause a failure, then none of the others should cause a failure. This property allows using only one value from each equivalence class as a representative for its set.

Equivalence class testing divides the test value domain into equivalence classes using contract conditions. Each test case selects one input value from each equivalence class. This approach is improved by boundary value selection of input values for numeric and date data, which appear at the boundaries of equivalence classes. For string data, such as names, and for other types of data, such as files, a set of input values representing each equivalence class should be manually prepared in advance with respect to the input contract and then test input values are selected randomly for each equivalence class. Thus, in our work, cause–effect testing, which generates test values from decision tables, is used to strengthen equivalence class testing. In the presented approach, causes are input conditions and effects are represented by actions. This proposed approach is presented in Algorithm 8, namely input contract-based test case generation algorithm, which derives test inputs from contract-supplemented ESG.

The input contract-based test case generation algorithm produces test values for each rule in the DT. The DT is represented with a data structure that contains the set of variables, the set of clauses along with its variable(s) and their equivalence classes, the set of actions and exceptions, and the set of rules wherein each rule is composed of a conjunction of clauses and conjunction of actions and exceptions.

For each rule, the function findTestInputValue is called. It attempts to find values for variables that satisfy the Boolean expression that is a special case of a *constraint satisfaction problem* [73] of the corresponding rule in the DT. The function solveCSP determines valid and invalid equivalence classes for each clause and searches the values that make the Boolean expression true. The runtime complexity of the whole algorithm mainly depends on this function, which has to be solved for each rule of the DT.

The algorithm of getAssignment within the function solveCSP starts by assigning a value to a single variable and extends the solution step by step with the other variables by assigning values. If a value assignment to the current variable is not possible due to previously selected values, the algorithm steps back and chooses next value from the set of boundary values for the current variable. This procedure is also called "simple backtracking." The proposed algorithm combines backtracking with the techniques "Arc Consistency Check" and "Minimum Remaining Values," see [73] for further information, to solve the given constraint satisfaction problem modeled by DT.

Algorithm 8. The Input Contract-Based Test Case Generation Algorithm

```
foreach event with DT do
    foreach rule in DT do
        findTestInputValue(DT, rulei)
    function findTestInputValue
    begin
```

```
    tc_inputs : Test_Case_Inputs
    set_clauses : LIST[<Clause, Variable, EquivalenceClasses >]
    b : BooleanExpr
    set_clauses ← getClauses(DT)
    b ← determineBooleanExpr(DT, rule)
    tc_inputs ← solveCSP(b, set_clauses)
    return tc_inputs
end
function solveCSP
begin
    assignment : LIST[<Variable , SelectedValue>]
    g ← getConstraintGraph(b, set_clauses)
    assignment ← getAssignment(g, assignment)
    return assignment
end
```

The runtime complexity for backtracking is given as $O(n*d)$, where n is the number of nodes for the corresponding constraint graph and d is the depth of the graph. The runtime complexity for the consistency check is given as $O(n^2 d^3)$ [73]. However, in practice the number of variables on a GUI is strictly limited due to usability restrictions.

Simultaneously, this also limits the number of corresponding constraints so that the runtime complexity of this algorithm is negligible. Furthermore, the search space for numerical values may be narrowed by considering only boundary values of equivalence classes. Finally, the function solveCSP returns test case inputs for a rule in the decision table. Resulting test cases contain test input values as well as expected results.

The development of test oracles, which automatically performs a pass/ fail evaluation of the test case, is an important issue in software testing. Developing such test oracles manually when writing test drivers is expensive and error-prone. Since our work proposes that assertions based on contracts can effectively be utilized as test oracles, the presented methodology is composed using different techniques to derive the oracles from the contracts in synchronization with the generation of test input values.

Fully automated testing requires automating the handling of oracles. In this case, evaluation of test results is straightforward due to the presence of contracts as specifications. Test cases are generated with expected test results automatically from the DT, which is constructed from input contracts. Since the test oracle in this approach uses executable input contracts by means of checking test case results, test outputs can be easily compared with expected test results. Thus, the test oracle in our work enables an automatic pass/fail evaluation of the test case. If the obtained results match the expected results, then the test case passes; otherwise it fails.

Analysis of the test case generation process reveals the fact that ESGs are to be transformed into one large model for test case generation. On the other hand, DTs could be consolidated, which results in the reduced number of rules. Both facts give some clues about the scalability of the presented approach. Transforming ESGs into one large model might complicate test case generation and the intuitive partitioning of SUT intended by the tester would be lost. Further test generation techniques are considerable, which make use of the intuitive partitioning of the tester to reduce and/or simplify test sequences and their generation, especially with regard to input contracts. The more input contracts exist, the costlier is their evaluation. This is due to the fact that adding just one single input contract doubles (in the worst case) the number of combinations of input contracts to be tested. Thus, further techniques to reduce the evaluation complexity of large sets of input contracts could be helpful, such as partitioning of input contracts that could be achieved by a hierarchical set of DTs.

The following questions are required to be answered in the future: (a) How much overhead does the presented approach impose on a tester of large software systems? (b) How far can a tester develop contracts for such an application and how long would it take? (c) How would a tester developing contracts for applications impact speed of execution during testing? Moreover, developing formal semantics for input contracts will have an important impact not only on the testing of GUIs but also on the design and implementation of GUIs. Definition of refinement and inclusion operations on contracts provides distinct means to express complex input behavior in terms of simpler behavior. Furthermore, a refinement enables specialization of contractual obligations and invariants of other contracts, whereas inclusion allows contracts to be composed from simpler subcontracts [74].

5. RATIONALIZATION AND AUTOMATION OF GUI TESTING

Techniques for modeling, analyzing, and testing GUIs are represented in the previous sections. This section categories and exemplifies tools for GUI testing that are available on the market. It is evident that this market changes frequently as these tools are to a great extent short-living. So, the critical reader may forgive if the authors have forgotten to name some important brands.

5.1 MBT Tools in General

MBT is defined as an approach to deriving executable tests from a given model of the SUT using several test selection criteria. The model is built either from requirements or from specifications of test model. The model should contain both input and expected output in order to be able to generate test oracles [75]. Thus, no model-based input generator and no test automation framework where test cases can be manually created or prerecorded [76] are considered to be MBT tools.

An MBT tool supports the test life cycle and interacts with other development phase elements as depicted in Fig. 16. Selected evaluation criteria are (i) Test Modeling—design of the test model derived from the SUT, (ii) Test Generation—strategy for deriving test cases, and (iii) Extensibility—integration possibilities with other tools via import/export interfaces and/or extending tool to different domains. The usability criterion has been intentionally omitted. Test generation algorithm is better covered elsewhere and omitted here.

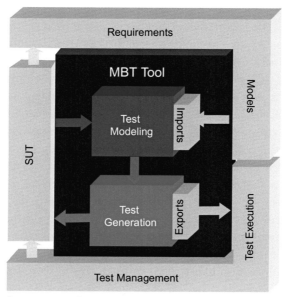

Fig. 16 MBT tool categories of interest.

MBT tools need to satisfy the following criteria: (i) Usage of a test model (not a system model) from where tests are derived. (ii) Automate test generation covering test input data and system behavior. (iii) The test model is represented by a formal modeling language.

5.2 Test Tools for GUIs

The term *test automation* for GUIs in industry implies automated test execution in most cases. There are many GUI test tools available [77,78] which we refer to as a selection of commercial as well as open-source tools. The test tool market however is changing rapidly. This makes it very difficult to discuss or compare specific tools which will be sustainable at the market. Thus, this section rather focuses on the needs from a practical and industrial perspective.

Test execution itself is mostly automated by *Capture/Replay* tools such as qftest [79] and jfcUnit [80] and is used for regression testing. More recently, *Capture/Replay* tools have been enhanced to visual GUI test tool using image recognition instead of GUI-object code or coordinates [81]. Test automation is more than just regression testing. Automation should be attempted for as many stages of the entire test process as possible. First, the test process has to be managed, which means that the test documentation, the different releases of test cases, and the test cases itself have to be maintained and kept consistent with the requirements they are associated with. In practice, this is the most neglected step. In the next step, the test cases have to be generated. This differs from a simple test script generator, which only allows defining test inputs and their corresponding outputs. When a specification represented as a model is used to automatically generate executable test, the method is referred to as *model-based* test generation. In contrast, in *data-driven* test generation [82], existing test cases of just plain templates are parameterized with different data the test has to be run with. Once the test cases are designed, tests have to be conducted on the SUT. Automation of this test step is supported by a wide variety of tools such as Conformiq Test Generator [83] and Smartesting CertifyIt [84] that support many popular programming languages. Finally, the test analysis step is left in which conducted tests and their outputs are evaluated. The results of the analysis are needed to fix the faults and to decide further tests.

The test automation has to keep through to the complete product life cycle. Kelly [85] divided the life cycle into six major stages, and for each stage, there is a checklist of questions. Depending on the responses, it is decided whether to automate the corresponding stage or not. However, the work [85] does not include necessary detail of how to set up test automation during the entire life cycle. Another work dealing with the issue about when a test should be automated is given in Ref. [86]. It assumes the intent of automated testing and does not take a decision on need for automation. Beyond this, it is still difficult to decide which tool fits best to the requirements. This is addressed in Ref. [87] providing a catalog of features to be compared when evaluating a GUI test tool.

5.3 Test Automation in Theory and Practice

This section describes test automation of the entire test process, divided into four parts as illustrated in Fig. 17, in the context of black-box testing of software applications containing a GUI. The objective of test automation research is to maximize automation in the test process. This means that existing models from the development are used to generate test cases (model driven) using suitable algorithms. The models have to contain the test inputs and the outputs to derive a fault model. This ensures that the tests are observable and the outputs can be compared with the expected ones described in the model. It may be observed in Fig. 17 that the test process steps are coordinated, i.e., the test results of one stage can be used as an input of the following stage. Therefore, the entire test process is automated as a single step.

Insights from industrial projects trying to adopt test automation reveal that test automation is treated for each step separately. This differs from theory which treats test automation as a single step. In Fig. 17, steps that still need some manual work are depicted as *semiautomated*. The test planning step is done manually and is accordingly depicted as *nonautomated*. In

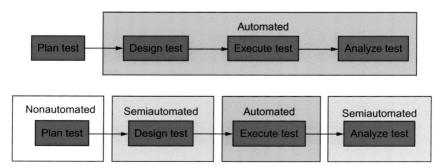

Fig. 17 Gap between test automation in theory and practice.

addition, automation is adopted in most cases too early which means without any preparation [86].

The most reasonable cause to implement test automation in practice is for repeating test cases. That is also the reason why until now regression testing is so widely used and test automation is almost limited to the test execution in industry. Regression testing just confirms that despite changes to the software, the tests previously run provide identical results now. In fact, this means that no new functions have been tested. So regression testing is not supposed to find new faults in the code. In theory, it is the methodology that reveals fewest bugs. However, the most Capture/Replay tools are not able to provide expected benefits.

5.4 Test Tool Requirements in Industry

For testing GUIs of software systems, powerful commercial tools, such as Ranorex [88], Squish [89], and HP Unified Functional Test (UFT) [90], are available. Nevertheless, numerous time-consuming and error-prone manual steps are still necessary to complete the test process. Still, not all test tools allow constructing templates which can be run for different data from a database. Also, not all testing tools nowadays support the use of stubs or wrapping. From the industrial point of view, there are many other problems concerning how to implement successful test automation. This section lists practical issues faced while attempting to introduce test automation for GUIs supported by Capture/Replay tools.

5.4.1 Dynamically Design Changes

Recorded objects of a GUI are identified by its unique properties. But these properties are static. This is an issue as only those objects which are available at the point of recording can be recognized by the tool. Otherwise, the objects have to be introduced into the tool afterward. But, this can be very time consuming because the system has to set in the right state where the elements are active. Another problem arises due to the changes of different objects from release to release. Although previously mentioned test tools (Ranorex, Squish, HP UFT) already provide to change the objects properties or to set regular expressions fitting the object, this has to be done manually and takes a lot of time to maintain. A significant problem is the case in which the object will change dynamically depending on the behavior of the user. For example, a conformed GUI depending on the user behavior affected by knowledge-based programs or learning software. There, it is impossible to predefine all possible cases where objects will change. These vulnerabilities can be eliminated by a model-based test generation, which maintains the model instead of repairing the test cases.

5.4.2 Failure Treatment

Failure treatment is rarely supported by the test tools. One problem observed when testing the GUI is that while the GUI objects are checked, the underlying system operation is not. Indeed, one can write a test scenario which tests if all fields have to be filled out and if the submission button is disabled after sending once. This restriction is enforced to prevent multiple entries. However, this does not prove that the system has really connected with the database and stored or changed the values. Therefore, the test tools have to be extended by a functionality to compare database values with the expected values. Another kind of fault appears when the application crashes. Then a fault can be detected, but the bug cannot be fixed, because there is no information on which test cases failed. The only way to find that out is by conducting the test cases manually. This is analogous to finding a zero point numerically. Hence, the advantage of executing the test scripts without being present is lost. The same holds for faults which prevent the process from continuing like open windows. It is assumed that each test case starts at the main window, i.e., from the same starting point. Therefore, the SUT has to be reset in its initial state from which each test case can be started. This is important if the test process should also provide test cases which are expected to fail. Pettichord mentions this special case of resetting the system in his work where he called it an "Error Recovering System" [91]. For bug fixing, it is also recommend having a function which can set the system in predefined system state from which the test can further run. Additionally, it may be necessary to have a memory map which is recorded during the test process. A worst-case scenario occurs when a test case crashes not only the application but also the testing tool or even the whole system. In this case the only chance to overcome the problem is to start the test execution from a remote machine which has to be provided by the test tool or even to support a test execution on a virtual machine.

5.5 ESG Test Suite Designer

Number of tools providing MBT is very limited. The ESG tool is a good example of MBT tool. The aim of this section is to explain ESG tool named ESG Test Suite Designer (ESG-TSD) [92] that provides necessary functions to develop ESG models and to generate test cases from ESG models.

Finite-state machines are models used in MBT. UML statechart is a way to represent finite-state machines. Ticket machine is used as a running example in this section. The running example has two ticket machines. Simple ticket machine lets users to buy a single-type ticket, whereas the complex ticket machine allows users to buy tickets by selecting ticket type as well as number of tickets. Their statecharts are given in Figs. 18 and 19, respectively.

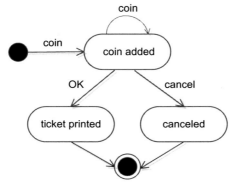

Fig. 18 Statechart for simple ticket machine [93].

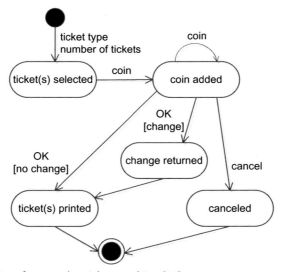

Fig. 19 Statechart for complex ticket machine [94].

Simple ticket machine shows coins inserted. User either inserts coins until the exact cost of ticket is reached and gets the ticket or quits and gets coins back. Complex ticket machine lets the user to select ticket type and number of tickets and then waits for the amount to be inserted. Once the inserted amount is equal to or more than the required amount, then the change is returned if there is and tickets are printed. The user can quit while inserting coins.

A statechart can be transformed into an ESG by presenting transitions between events that cause state transitions in statecharts. ESG for simple ticket machine is modeled using ESG-TSD and shown in Fig. 20, which shows valid event sequences for simple ticket machine. Complete event sequences, which are given in Fig. 21, are generated by pressing wheel icon on the icon tab of ESG-TSD. The output shown in Fig. 21 also presents the time elapsed in producing complete event sequences.

ESG for complex ticket machine is shown in Fig. 22. When wheel icon is pressed, it generates three complete event sequences as follows:
- "[, select ticket type, enter number of tickets, insert coin, print ticket(s),]"
- "[, select ticket type, enter number of tickets, insert coin, cancel, return money,]"
- "[, select ticket type, enter number of tickets, insert coin, insert coin, print ticket(s), return change,]"

ESG-TSD also produces FCESs, which are test cases for negative testing. For instance, "[, select ticket type, insert coin,]" is a faulty complete event sequence of length 2 for complex ticket machine. ESG-TSD generates 15 FCESs for simple ticket machine and 48 FCESs for complex ticket machine.

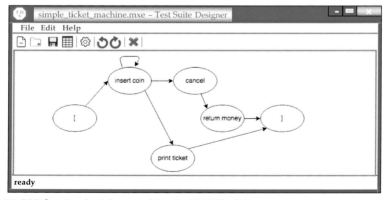

Fig. 20 ESG for simple ticket machine in ESG-TSD [95].

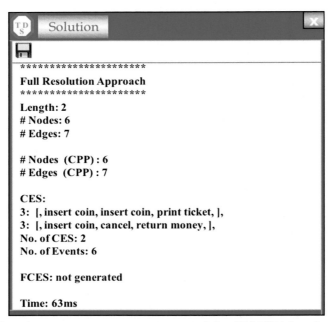

Fig. 21 Complete event sequences generated by ESG-TSD for simple ticket machine [95].

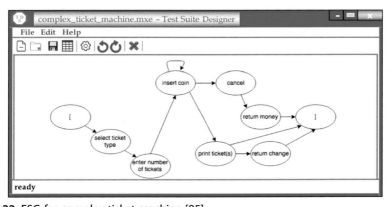

Fig. 22 ESG for complex ticket machine [95].

DT-supplemented ESGs are supported by the ESG-TSD. As explained earlier, DTs help to reduce complexity of ESGs by representing conditional event transitions in DTs within sub-ESGs. DT-supplemented ESG for complex ticket machine is given in Fig. 23. The difference between Figs. 22 and 23 is that "print ticket(s)" and "return change" events are encapsulated in

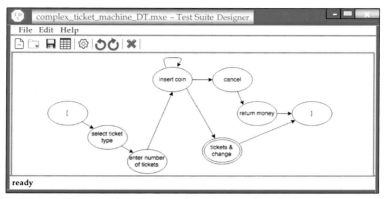

Fig. 23 DT- supplemented ESG for complex ticket machine [95].

"tickets & change" sub-ESG represented using double circle. In this sub-ESG after printing ticket(s), if there is no change to be returned, pseudo-exit is executed, whereas if there is a change, after printing ticket(s) first change is returned and then pseudo-exit is executed. This is an example of how to simplify ESG models.

6. CONCLUSIONS

GUIs are likely to continue playing a role in HCIs. The question is how far new modes of interaction may make the field obsolete. In other words, can we reuse and generalize some of the things we have learned with GUI testing and apply them to other types of software?

Before wrapping up and summarizing this chapter, the authors briefly discuss on where they see the field going.

6.1 Peer Into the Future

6.1.1 Changes in Look and Feel of UIs Will Come, but What About Their Modeling?

GUIs have been important parts of many software applications to provide interactions between the user and the system. Recent advances suggest that UIs are likely to change form in future. Although classical GUI elements, such as buttons and menus, will probably be there for at least some time, the way that users interact with systems is expected to change substantially. It will not be surprising when interacting with systems using additional types of triggers such as sounds, hand/body/face/eye gestures, and brain waves becomes a common thing. The interaction methods supported by UIs have been getting enriched by such technological advances. Therefore, the approaches that do not have right level of abstraction or that are strongly dependent on certain GUI properties or components are expected to be harder to reuse or adapt in future when the changes reach at a considerable level. However, event-based approaches as discussed in Section 2 can still be useful if the user interacts with the system in a way that can be formulated in terms of discrete stimuli. It is always possible that semantic extensions will be required to capture different types of behaviors, descriptions, or problems; however, the formal basis will be the same.

On the other hand, the approaches proposed for GUI testing can also be used for other types of software applications with certain adjustments depending on the application area. As mentioned earlier, if user–system interactions can be formulated in the form of events, most of the approaches proposed for GUI testing can be used. For example, event-based models such as ESGs are used for testing of web service compositions, and, also, there are formalisms which can be alternative/complementary to the finite-state-machine-based approaches in fault-based testing.

6.1.2 Contracts Will Become More Precise and Look Different

Further test generation techniques, which make use of the intuitive approaches to reduce and/or simplify test sequences and their generation, are considerable for contract-based GUI testing. The more GUI contracts exist, the costlier is their evaluation. This is because adding just one single contract doubles (in the worst case) the number of combinations of contracts to be tested. Thus, further techniques to reduce the evaluation complexity of large sets of GUI contracts can be helpful, such as partitioning of contracts that can be achieved by a hierarchical set of DTs, if GUI contracts are represented with DTs.

Another path for future work is to introduce formal semantics for GUI contracts, which may include refinement and inclusion operations on contracts. These operations aim to provide distinct means to express complex GUI behavior in terms of simpler behavior. Furthermore, a refinement enables specialization of contractual obligations and invariants of other contracts, whereas inclusion allows contracts to be composed from simpler contracts. With these contract operations, an approach for combining GUI components and testing them as a single unit can be defined and implemented. More formalism for GUI contracts is necessary to systematically discuss and justify the differences between GUI contracts and classical Meyer contracts with respect to preconditions, postconditions, invariants, and inheritance mechanism.

Such formalism on GUI contracts may end up with a formal contract language for specification of GUIs with composition and inheritance mechanisms. Contracts written in this language can be converted to test cases and test oracles can be built automatically.

6.1.3 What About Automation of GUI Testing?

MBT has been proven to bring advantage in terms of test effectiveness and test efficiency when testing GUIs. However GUIs have been evolving over time from systems where it was easily possible to test merely every interface in the past to more complicated systems. Today's challenge is to know what to test for in GUIs rather than how to test as not everything is testable anymore. Future GUIs are heading toward more intelligent systems with changing feedback behavior over time not only adapting to the user behavior but even predicting the user's behavior. Furthermore, GUIs will integrate voice interfaces utilizing speech processing (such as apple's Siri). The results of these more intelligent user interfaces will become unpredictable and bring the challenge how to test them if we don't know what to test for.

MBT tends to generate a large amount of test cases which will become impractical even if executed automatically for future GUIs. On the contrary side testing needs to withstand the coverage of the exponential growth of GUIs comprising more and more features. Thus, testing GUIs will require even more emphasis to optimize and minimize test suites in the future. The goal is to find coverage of GUI interactions and events by test instances of test inputs and test sequences with high defect detection likelihood. This would also mean that traditional test coverage criteria are probably not adequate any longer and have to be replaced by a more property- or situation-based test input generation. In addition, the automation of the test oracle would need a heuristic approach when testing complex GUIs to overcome their characteristics of unpredictable test outcomes. This would tend toward an oracle that would find suspicious deviations and would need to define some new qualitative measurements.

6.2 Summary

This chapter reviewed existing work on model-based GUI testing. Both the desirable and undesirable features of the system to be developed have been unified, leading to a holistic approach to testing. Notions of mutation analysis and testing have been used to refine this holistic view.

Event-based modeling has been favored as it is broadly accepted. Due to the use of models that can be represented as graphs, application of sound mathematical methods is enabled. Results and algorithms have been borrowed from graph theory, automata theory, and formal languages to construct test cases, optimize test suites, etc. Modeling with ESGs has been exemplarily used without any loss of generality, because also other graph-based models can be adapted for enabling the application of such formal methods.

The focus was on modeling and test case construction that covered also optimization. An important issue of testing in the practice is the quantification of the test cases, that is, assigning real and symbolic values to them. This aspect has been studied involving contract-based testing, which entails augmenting the graphs by decision tables.

Because of the limitation of the expected size of a chapter not all aspects of GUI testing could be discussed, for example, test prioritization, or considering other semantic features than "follows" relation in ESG modeling, for example, concurrency or causality. Follow-on limitations can be seen in the definition and usage of fault and coverage semantics. Consequently, there are severe theoretical barriers, necessitating further research to extend and generalize the introduced ideas and techniques, mostly caused by the explosion of states when taking, for example, concurrency into account [22].

Nevertheless, the authors hope that further research will enable more and comprehensive adoption of the approaches introduced in the practice.

REFERENCES

[1] J.A. Whittaker, Software's invisible users, Softw. IEEE 18 (3) (2001) 84–88.
[2] L. White, H. Almezen, Generating test cases for GUI responsibilities using complete interaction sequences, in: Proceedings of the International Symposium on Software Reliability Engineering ISSRE, IEEE Computer Society Press, Washington, DC, USA, 2000, pp. 110–119.
[3] B. Korel, Automated test data generation for programs with procedures, Proceedings of ISSTA'96, (1996) 209–215.
[4] A.M. Memon, M.E. Pollack, M.L. Soffa, Hierarchical GUI test case generation using automated planning, IEEE Trans. Softw. Eng. 27 (2) (2001) 144–155.
[5] D. Hamlet, Foundation of software testing: dependability theory, Proceedings of ISSTA'96, (1994) 84–91.
[6] M.A. Friedman, J. Voas, Software Assessment, John Wiley & Sons, New York, 1995.
[7] F. Belli, Finite-state testing and analysis of graphical user interfaces, in: Proceedings of the 12th International Symposium on Software Reliability Engineering (ISSRE), IEEE Computer Society, Washington, DC, USA, 2001, pp. 34–43.

[8] A.M. Memon, M.L. Soffa, M.E. Pollack, Coverage criteria for GUI testing, in: Proceedings of the 8th European Software Engineering Conference Held Jointly With 9th ACM SIGSOFT International Symposium on Foundations of Software Engineering (ESEC/FSE-9), ACM, New York, NY, USA, 2001, pp. 256–267.

[9] F. Belli, M. Beyazıt, Using regular grammars for event-based testing, in: S. Konstantinidis (Ed.), Proceedings of the 18th International Conference on Implementation and Application of Automata (CIAA), Lecture Notes in Computer Science, vol. 7982, Springer, Heidelberg, 2013, pp. 48–59.

[10] F. Belli, M. Beyazıt, Exploiting model morphology for event-based testing, IEEE Trans. Softw. Eng. 41 (2) (2015) 113–134.

[11] L. White, H. Almezen, User-based testing of GUI sequences and their interactions, in: Proceedings of the 12th International Symposium on Software Reliability Engineering (ISSRE), IEEE Computer Society, Washington, DC, USA, 2001, pp. 54–63.

[12] F. Belli, M. Beyazıt, A.T. Endo, A. Mathur, A. Simao, Fault domain-based testing in imperfect situations—a heuristic approach and case studies, Softw. Q. J. 23 (3) (2015) 423–452.

[13] R.K. Shehady, D.P. Siewiorek, A method to automate user Interface testing using variable finite state machines, in: Proceedings of the 27th Annual International Symposium on Fault-Tolerant Computing (FTCS), IEEE Computer Society, Washington, DC, USA, 1997, pp. 80–88.

[14] A.C.R. Paiva, N. Tillmann, J.C.P. Faria, R.F.A.M. Vidal, Modeling and testing hierarchical GUIs, The 12th International Workshop on Abstract State Machines (ASM), (2005) 8–11.

[15] F. Belli, C.J. Budnik, A. Hollmann, Holistic testing of interactive systems using statecharts, J. Math. Comput. Teleinform. (AMCT) 1 (3) (2005) 54–64.

[16] F. Belli, A. Hollmann, Test generation and minimization with 'basic' statecharts, in: Proceedings of the 23rd ACM Symposium on Applied Computing (SAC), ACM, New York, NY, USA, 2008, pp. 718–723.

[17] F. Belli, M. Beyazit, T. Takagi, Z. Furukawa, Mutation testing of go-back functions based on pushdown automata, in: Proceedings of the 2011 IEEE 4th International Conference on Software Testing, Verification and Validation (ICST), IEEE Computer Society, Washington, DC, USA, 2011, pp. 249–258.

[18] F. Belli, M. Beyazıt, T. Takagi, Z. Furukawa, Model-based mutation testing using pushdown automata, IEICE Trans. Inf. Syst. E95-D (9) (2012) 2211–2218.

[19] Q. Xie, A.M. Memon, Using a pilot study to derive a GUI model for automated testing, ACM Trans. Softw. Eng. Methodol. 18 (2) (2008) 1–35.

[20] F. Belli, M. Beyazıt, N. Güler, Event-oriented, model-based GUI testing and reliability assessment—approach and case study, in: A. Memon (Ed.), Advances in Computers, vol. 85, Elsevier, Amsterdam, Netherlands, 2012, pp. 277–326.

[21] Private Communication, The idea of ESG extension, mainly reflecting Professor Ina Schieferdecker's idea, bases on early, unpublished discussions between her and Fevzi Belli, 2006.

[22] F. Belli, M. Beyazit, A. Memon, Testing is an event-centric activity, in: Proceedings of the 6th International Conference on Software Security and Reliability (SERE-C 2012), IEEE Computer Society, Washington, DC, USA, 2012, pp. 198–206.

[23] A.M. Memon, I. Banerjee, A. Nagarajan, GUI ripping: reverse engineering of graphical user interfaces for testing, in: Proceedings of the 10th Working Conference on Reverse Engineering (WCRE), IEEE Computer Society, Washington, DC, USA, 2003, pp. 260–269.

[24] A.M. Memon, Q. Xie, Studying the fault-detection effectiveness of GUI test cases for rapidly evolving software, IEEE Trans. Softw. Eng. 31 (10) (2005) 884–896.

[25] X. Yuan, A.M. Memon, Using GUI run-time state as feedback to generate test cases, in: Proceedings of the 29th International Conference on Software Engineering (ICSE), IEEE, Washington, DC, USA, 2007, pp. 396–405.

[26] P. Brooks, A.M. Memon, Automated GUI Testing guided by usage profiles, in: Proceedings of the 22nd IEEE/ACM International Conference on Automated Software Engineering (ASE), ACM, New York, NY, USA, 2007, pp. 333–342.

[27] F. Belli, M. Beyazit, A formal framework for mutation testing, in: Proceedings of the 2010 4th International Conference on Secure Software Integration and Reliability Improvement (SSIRI), IEEE Computer Society, Washington, DC, USA, 2010, pp. 121–130.

[28] E.F. Moore, Gedanken experiments on sequential machines, automata studies, in: C.E. Shannon, J. McCarthy (Eds.), Annals of Mathematical Studies, vol. 34, Princeton University Press, Princeton, NJ, USA, 1956, pp. 129–153.

[29] G.H. Mealy, A method for synthesizing sequential circuits, Bell Syst. Tech. J. 34 (1955) 1045–1079.

[30] R.G. Hamlet, Testing programs with the aid of a compiler, IEEE Trans. Softw. Eng. SE-3 (1977) 279–290.

[31] R.A. DeMillo, R.J. Lipton, F.G. Sayward, Hints on test data selection: help for the practicing programmer, IEEE Comput. 11 (1978) 34–41.

[32] T.A. Budd, A.S. Gopal, Program testing by specification mutation, Comput. Lang. 10 (1) (1985) 63–73.

[33] M. Beyazit, Exploiting model morphology for event-based testing, PhD Thesis, University of Paderborn, 2014.

[34] F. Belli, C.J. Budnik, W.E. Wong, Basic operations for generating behavioral mutants, in: Proceedings of the 2nd Workshop on Mutation Analysis (MUTATION), IEEE Computer Society, Washington, DC, USA, 2006, pp. 9–18.

[35] M.K. Kwan, Graphic programming using odd or even points, Chin. Math. 1 (3) (1962) 273–277.

[36] J. Edmonds, E.L. Johnson, Matching, Euler tours and the Chinese postman, Math. Program. 5 (1) (1973) 88–124.

[37] A. Aho, A. Dahbura, D. Lee, M. Uyar, An optimization technique for protocol conformance test generation based on UIO sequences and rural Chinese postman tours, IEEE Trans. Commun. 39 (1991) 1604–1615.

[38] F. Belli, M. Beyazit, N. Güler, Event-based GUI testing and reliability assessment techniques—an experimental insight and preliminary results, in: Proceedings of the 2011 IEEE 4th International Conference on Software Testing, Verification and Validation Workshops (ICSTW 2011/TESTBEDS), IEEE Computer Society, Washington, DC, USA, 2011, pp. 212–221.

[39] P. Purdom, A sentence generator for testing parsers, BIT Numer. Math. 12 (3) (1972) 366–375.

[40] B.A. Malloy, J.F. Power, A top-down presentation of Purdom's sentence-generation algorithm, Technical Report, NUIM-CS-TR-2005-04, National University of Ireland, Maynooth, Ireland, 2005.

[41] L. Zheng, D. Wu, A sentence generation algorithm for testing grammars, Proceedings of the 33rd International Computer Software and Applications Conference (COMPSAC), vol. 1, IEEE Computer Society, Washington, DC, USA, 2009, pp. 130–135.

[42] A. M. Memon, M. E. Pollack and M. L. Soffa, Automated test oracles for GUIs, SIGSOFT, (2000) 30–39.

[43] M. Marré, A. Bertolino, Using spanning sets for coverage testing, IEEE Trans. Softw. Eng. 29 (11) (2003) 974–984.

[44] A. Gargantini, C. Heitmeyer, Using model checking to generate tests from requirements specification, Proceeding of the ESEC/FSE. ACM SIGSOFT, New York, NY, USA, 1999, pp. 146–162.

[45] D.B. West, Introduction to Graph Theory, Prentice Hall, Upper Saddle River, NJ, USA, 1996.

[46] H. Thimbleby, The Directed Chinese Postman Problem, School of Computing Science, Middlesex University, London, 2003.

[47] W. Edger, A. Dijkstra, Note on two problems in connexion with graphs, J. Numer. Math. 1 (1959) 269–271.

[48] F. Belli, C.J. Budnik, Test minimization for human–computer interaction, Appl. Intell. 26 (2) (2007) 161–174.

[49] R. K. Shehady and D. P. Siewiorek, A method to automate user interface testing using finite state machines, in Proceedings of the International Symposium on Fault-Tolerant Computing, FTCS-27, (1997) 80–88.

[50] Christof J. Budnik, Fevzi Belli, Axel Hollmann, Structural feature extraction for GUI test enhancement, ICST Workshops, (2009) 255–262.

[51] C. Wohlin, P. Runeson, Experimentation in Software Engineering—An Introduction, Kluwer Academic Publishers, Dordrecht, Netherlands, 2000.

[52] F. Belli, N. Nissanke, C.J. Budnik, A holistic, event-based approach to modeling, analysis and testing of system vulnerabilities, Technical Report TR 2004/7, University of Paderborn, Paderborn, 2004.

[53] F. Belli, C.J. Budnik, L. White, Event-based modelling, analysis and testing of user interactions: approach and case study, Softw. Test. Verif. Reliab. 16 (1) (2006) 3–32.

[54] B. Meyer, Applying design by contract, Computer (Long. Beach. Calif.). 25 (10) (1992) 40–51.

[55] J.M. Jazequel, B. Meyer, Design by contract: the lessons of Ariane, Computer 30 (1) (1997) 129–130.

[56] T. Tuglular, C.A. Muftuoglu, F. Belli, M. Linschulte, Model-based contract testing of graphical user interfaces, IEICE Trans. Inf. Syst. E98.D (7) (2015) 1297–1305.

[57] C.D.T. Cicalese, S. Rotenstreich, Behavioral specification of distributed software component interfaces, Computer (Long. Beach. Calif.). 32 (7) (1999) 46–53.

[58] W. Zheng, G. Bundell, Test by contract for UML-based software component testing, in Proceedings of IEEE International Symposium on Computer Science and Its Applications, Washington, DC, USA, (2008) 377–382.

[59] I. Ciupa, A. Leitner, Automatic testing based on design by contract, in Proceedings of Net. ObjectDays (6th Annual International Conference on Object-Oriented and Internet-Based Technologies, Concepts, and Applications for a Networked World), (2005) 545–557.

[60] L.C. Briand, Y. Labiche, H. Sun, Investigating the use of analysis contracts to improve the testability of object-oriented code, Softw. Pract. Exp. 33 (7) (2003) 637–672.

[61] B.K. Aichernig, Contract-based mutation testing in the refinement calculus, Electron. Notes Theor. Comput. Sci. 70 (3) (2002) 281.

[62] P. Madsen, in: Testing by contract—combining unit testing and design by contract, The Tenth Nordic Workshop on Programming and Software Development Tools and Techniques, 2002. see also: http://www.itu.dk/people/kasper/NWPER2002/papers/madsen.pdf.

[63] P. Guerreiro, Simple support for design by contract in C++, In Proceedings of the 39th International Conference and Exhibition on Technology of Object-Oriented Languages and Systems, July 29–August 03 2001, IEEE, Washington, DC.

[64] R. Plösch, Design by contract for python, IEEE Proceedings of the Joint Asia Pacific Software Engineering Conference (APSEC97/ICSC97), HongKong, December 2–5, 1997.

[65] R. Heckel, M. Lohmann, Towards contract-based testing of web services, Electron. Notes Theor. Comput. Sci. 116 (2005) 145–156.

[66] G. Dai, X. Bai, Y. Wang, F. Dai, Contract-based testing for web services, Computer Software and Applications Conference, COMPSAC 2007. 31st Annual International, vol. 1, (2007) 517–526.

[67] E. Valentini, G. Fliess, E. Haselwanter, A framework for efficient contract-based testing of software components, 29th Annual International Computer Software and Applications Conference (COMPSAC'05), vol. 1, (2005) 219–222.

[68] D. Xu, W. Xu, M. Tu, Automated generation of integration test sequences from logical contracts, Computer Software and Applications Conference Workshops (COMPSACW), 2014 IEEE 38th International. (2014) 632–637.

[69] IEEE Std 610.12-1990, IEEE Standard Glossary of Software Engineering Terminology, 1990.

[70] B. Baudry, Y. Le Traon, and J. Jezequel, Robustness and diagnosability of OO systems designed by contracts, Seventh International Software Metrics Symposium, METRICS 2001, Proceedings, (2001) 272–284.

[71] T. Tuglular, C. A. Muftuoglu, F. Belli, M. Linschulte, Event-based input validation using design-by-contract patterns, ISSRE 2009, 20th International Symposium on Software Reliability Engineering, Mysuru, Karnataka, India, (2009) 195–204.

[72] T. Tuglular, F. Belli, M. Linschulte, Input contract testing of graphical user interfaces, Int. J. Softw. Eng. Knowl. Eng. 26 (02) (2016) 183–215.

[73] S. J. Russell, P. Norvig, J. F. Canny, J. M. Malik, D. D. Edwards, Artificial Intelligence: A Modern Approach, vol. 2, Prentice Hall, Englewood Cliffs, USA, 1995.

[74] R. Helm, I.M. Holland, D. Gangopadhyay, Contracts: specifying behavioral compositions in object-oriented systems, ACM 25 (10) (1990) 169–180.

[75] A. Pretschner, J. Philipps, 10 methodological issues in model-based testing, model-based testing of reactive systems, Lect. Notes in Comput. Sci. 3472 (2005) 281–291.

[76] A. Hartmann, AGEDIS Model Based Test Generation Tools, AGEDIS Consortium, 2002.

[77] Wiki Comparison of GUI Testing Tools, https://en.wikipedia.org/wiki/Comparison_of_GUI_testing_tools, July 2014.

[78] Gartner Report on Software Test Automation Tools, 2016.

[79] qftestJUI Homepage, At URL: http://www.qfs.de.

[80] jfcUnit Homepage, At URL: http://jfcunit.sourceforge.net.

[81] Sekuli Script Homepage, At URL: http://www.sikuli.org/.

[82] ISO/IEC/IEEE 29119 Testing Standard: http://www.softwaretestingstandard.org/index.php.

[83] Conformiq Software: Conformiq Test Generator Homepage, At URL: http://www.conformiq.com.

[84] Smarttesting CertifyIt Homepage, At URL: http://www.smartesting.com/en/certifyit/.

[85] D. Kelly, Software Test Automation and the Product Life Cycle, 13(10), MacTech, Westlake Village, CA, USA, 1999.

[86] B. Marick, When should a test be automated, http://www.exampler.com/testing-com/writings/automate.pdf, 1998.

[87] E. Hendrickson, Making the right choice: the features you need in a gui test automation tool, STQE Mag. 3 (1999) 20–25.

[88] Ranorex Test Automation Homepage, At URL: http://www.ranorex.com.

[89] Froglogic Squish Homepage, At URL:https://www.froglogic.com/squish/.

[90] HP Unified Functional Testing (UFT) Homepage, At URL: http://www8.hp.com/de/de/software-solutions/unified-functional-automated-testing/.

[91] B. Pettichord, Success With Test Automation, Quality Week, San Francisco, CA, USA, 2001. https://www.prismnet.com/~wazmo/succpap.htm.

[92] ESG Test Suite Designer (ESG-TSD) homepage, At URL: http://download.ivknet.de/.

[93] M. Hübner, I. Philippow, and M. Riebisch, Statistical usage testing based on UML, Proceedings of the 7th World Multiconferences on Systemics, Cybernetics and Informatics, Orlando, FL, USA, July 27, 2003.

[94] R.K. Swain, P.K. Behera, D.P. Mohapatra, Minimal testcase generation for object-oriented software with state charts, Int. J. Softw. Eng. Appl. 3 (4) (2012) 39–59.

[95] F. Belli, M. Linschulte, T. Tuglular, Karar Tablosu Destekli Olay Sıra Çizgeleri Temelli Sınama Durum Üretim Aracı, Proceedings of the 10th Turkish National Software Engineering Symposium, Canakkale, Turkey, (2016) 408–413.

ABOUT THE AUTHORS

Dr. Belli is a professor emeritus of software engineering at University of Paderborn and at Izmir Institute of Technology. He has more than 35 years' experience in research, development and teaching software engineering, validation & verification, fault tolerance, and quality assurance. He started as a programmer in the aircraft industry and wrote programs to create simulation environments and to validate safety critical features. In 1983, he was awarded a professorship at the University of Applied Sciences in Bremerhaven; in 1989 he moved to the University of Paderborn. Dr. Belli was also, for many years, a faculty member of the University of Maryland, College Park, European Division. During 2002 and 2003, he was founding chair of the Computer Science Department at the University of Economics in Izmir, Turkey. Dr. Belli has an interest and experience in software reliability/fault tolerance, model-based testing, and test automation.

Mutlu Beyazit received the BSc degree in computer engineering and the MSc degree in computer software from Izmir Institute of Technology in 2005 and 2008, respectively. In 2014, he received the PhD degree from the University of Paderborn. He was a research assistant at Izmir Institute of Technology from 2005 to 2008 and at the University of Paderborn from 2009 to 2013. Since 2014, he has been with Yaşar University. His current interests include model-based testing.

Dr. Christof J. Budnik is a Senior Key Expert Engineer at the Architecture and Verification of Intelligent Systems Research Group of Siemens Corporation, Corporate Technology in Princeton, NJ. He leads research and business projects in several industrial domains, striving for advanced technology solutions exploiting artificial intelligence to improve and invent test approaches. Before joining Siemens he was the head of software quality for a German company in the smart card business.

Christof obtained his PhD in Electrical Engineering 2006 from the University of Paderborn, Germany, on event-based testing. He is author of many published contributions at several journals and conferences comprising his research interests in the areas of model-based testing, mutation testing, test automation, and formal verification. He organizes workshops on software testing and serves as guest reviewer for selected journals and is pc member of many international conferences.

Tugkan Tuglular received the BS, MS, and PhD degrees in Computer Engineering from Ege University, Turkey in 1993, 1995, and 1999, respectively. He worked as a research associate at Purdue University from 1996 to 1998. He has been with Izmir Institute of Technology since 2000. His current research interests include model-based testing and model-based software development.

Printed in the United States
By Bookmasters